Genes, Categories, and Species

Genes, Categories, and Species

The Evolutionary and Cognitive Causes of the Species Problem

JODY HEY

OXFORD

UNIVERSITY PRESS

2001

OXFORD
UNIVERSITY PRESS

Oxford   New York

Athens   Auckland   Bangkok   Bogotá   Buenos Aires   Cape Town
Chennai   Dar es Salaam   Delhi   Florence   Hong Kong   Istanbul   Karachi
Kolkata   Kuala Lumpur   Madrid   Melbourne   Mexico City   Mumbai   Nairobi
Paris   São Paulo   Shanghai   Singapore   Taipei   Tokyo   Toronto   Warsaw

and associated companies in
Berlin   Ibadan

Copyright © 2001 by Oxford University Press, Inc.

Published by Oxford University Press, Inc.
198 Madison Avenue, New York, New York 10016

Oxford is a registered trademark of Oxford University Press.

Library of Congress Cataloging-in-Publication Data
Hey, Jody.
Genes, categories, and species : the evolutionary and cognitive causes of
the species problem / Jody Hey.
p.   cm.
Includes bibliographical references.
ISBN 0-19-514477-5
1. Species.   I. Title.
QH83 .H53   2001
576.8'6—dc21   00-045655

1 3 5 7 9 8 6 4 2

Printed in the United States of America
on acid-free paper

To Kathy

PREFACE

Our lives are immersed in biological diversity. We are organisms of one species, and we watch, study, play with, put to work, grow, kill, eat, and suffer diseases caused by organisms of other species. Of course, our senses and thoughts are highly tuned to the detection and presence of other organisms, and in various ways we are experts at identifying kinds of organisms. As children, our first words are often names for kinds of organisms, and as adults our occupations are often grounded in a knowledge base of organismal diversity (e.g., gardeners, cooks, health care specialists). Biologists, too, must have a broad and deep knowledge of kinds of organisms. But in the very profession of the study of biological diversity, we find a great irony. Biologists are frequently uncertain, and often cannot agree on, how to identify species and on how to define the word SPECIES. These two uncertainties are closely tied to one another, and together they constitute what is known as the species problem.

The sticking points of the species problem, the frequent uncertainty over identifying species and defining SPECIES, do not form a conventional scientific puzzle. Though many biologists are aware of the problem, perhaps even to the point of despair that another book should be written on the topic, they do not address the problem as they do most questions. They do not ask, "What new information do we need to solve the problem?" The absence of this question from species debates is a clue that resolution of the problem is not to be found either in a finer description of biological diversity or in a more finely crafted definition of the word SPECIES.

Genes, Categories, and Species describes the pursuit of what is left to inquire of, and that is the idea that biologists are somehow predisposed to poorly understand and poorly communicate about species. That pursuit draws on discoveries that have been made in several different fields where researchers have inquired of the ways that people use language to describe the natural world. It was great fun to find a thread of reasoning that coursed through the diverse fields of biology, philosophy, anthropology, and psychology; and that leads to an explanation of our troubles. That explanation is not a species problem solution of the sort that some have hoped for, and it does not simply dispel our uncertainties about species. But if it is correct, then the explanation can be used by biologists to overcome some of the ways that we suffer the species problem.

The crux of the species problem is the way that people devise and rely on categories. We are compelled to construct and use categories of organisms, and we are compelled to understand biological diversity, and these two motives are frequently incommensurable. It is this conflict that drives our misguided behaviors with respect to species—misguided in hindsight of understanding the species problem. One such behavior is our penchant for devising pithy definitions of SPECIES. Dozens have been proposed in recent decades. I have participated in this, and not a few biologists have seen one of their major goals to be the crafting of just the right short series of words that will lay the species problem to rest. Another such behavior is the counting of real species. Biologists are steadily reporting new larger counts of the numbers of species that they suppose are really out there. Yet we do so while bearing the knowledge that many real species are not distinct and not countable. The devising of crisp species definitions and the counting of species, as well as other symptoms of the species problem, are misplaced efforts at making species tractable. Though understandable, and reasonable in some lights, they appear as compulsions when they are lit by an understanding of the species problem.

I am far from free of these compulsions, and as an evolutionary biologist, I am a regular sufferer of the species problem. This familiarity lead me to write this book while envisioning other evolutionary biologists as readers. However, no part of the explanation seemed to require a strong reliance on the specialized lexicons that arise in biology and in species debates. I also hoped that this work would be of interest to investigators in those other fields from which I have drawn pieces of the argument. For these reasons, I have tried to write words that would be found accessible to biologists and also to readers with just a basic understanding of modern biology. The text may in places seem overly simplistic to a professional biologist, particularly one who studies species in one fashion or another. I hope that expert readers bear with me when it seems too simple. The reduced perspective that I've tried to maintain throughout the book is in no part intended to permit mistakes or to avoid necessary complex issues.

A background in genetics or evolutionary biology is helpful for chapters 6, 7, and 10. Those chapters are core components of the main argument of the book, for they address the evolutionary side of the question. However, the overall thesis is a synthetic one that draws from several different fields, and it should be possible to understand much of the larger thesis even without a full grasp of the content of those chapters.

This book has three main sections. Chapters 1–3 introduce the species problem, the method of attack, and the necessary background for that attack. The method begins with a question, but it is not the conventional question of "What are species?" In place of this familiar question is one that is unconventional in the context of the species problem. Because the question is new, the method of resolution lies somewhat outside the main tradition of inquiry on the species problem. Indeed, the method is indirect—though hopefully a reader will appreciate the detective story aspect that I felt while following it and writing about it. The first three chapters also describe the need for that indirect approach and lay the groundwork for it. The second part of the book (Chapters 4–9) contains that indirect attack on the species problem and thus constitutes the primary discovery phase of the project. By the end of part two, there is a description of the causes of the problem as well as an evolutionary hypothesis of how we came to have the problem. The third section, that makes up the remaining chapters, reviews the ways that the species problem plays out for working biologists, and it explains biologists' behaviors in the light of an understanding of the species problem. The final chapter explains some recommendations. Although they may not be easy to follow, they are neither utopian nor radical.

Words as Symbols

Regrettably, I must devote a few words to a tedious matter and ask readers to take notice of the way that some words are presented in small capital letters. The book is partly about words as symbols for things, and it may not make sense, literally, without some explanation of how underlining has been used to alter the meanings of words. Species are one major subject of the book, but to write about them and about our consternation over them, it was absolutely necessary that I be able to refer to the word for species, purely and simply, without its usual meaning. Thus much of the book is about a word, SPECIES, and by the capital letters I have indicated that the word does not refer to anything, but rather is to be taken simply as a piece of text. As similar as the word SPECIES may seem to its noncapitalized counterpart, it would be hard to exaggerate the difference in meaning between the two cases. As it appears in the title of the book, without capital letters, the word means some kind of group of organisms, but when it is presented in capital letters it is simply the symbol that we use in written text to refer to those sorts of things. Throughout the book, whenever a word or phrase is in capital letters, that word or phrase should be interpreted as a mere symbol. In this way, by using words in their usual way and by sometimes using capital letters and drawing attention to them as symbols, it is possible to write directly about specific words.

Quotation marks are another means by which writers sometimes strip the usual meaning from a word. Unfortunately, the precise meaning that remains when quotation marks are used is not always clear. If they flank a word, without explanation, it may not be clear whether the word should be taken to mean a symbol, a quotation, or as an uncertain signpost to the reader that the usual meaning does not apply. To avoid confusion, I have used capital letters rather than quotation marks to refer to words directly, as symbols.

Apologies

On its way toward the explanation of the species problem, this book draws on insights from several different fields. Later, when drawing from that explanation, the book turns to a number of historically problematic questions faced by biologists who study species. Both coming and going, the purpose of the book takes it over varied terrain. Yet the text touches on but a portion of the diverse subject matter that might be considered apropos of the species problem. I found, while writing, what many other students of the species problem must have found, that the diversity of the subject matter is a daunting distraction. Certainly, I found it difficult at times to pull together the findings that ultimately formed this book, which often relied on diverse sources, while avoiding being drawn and quartered by the many, many tangents that presented themselves. My recourse was just a deliberate focus on purpose and succinctness. Although I recognize that I have not been as focused and brief as we might wish, I also know that I could not have articulated the points demanded by the argument, and I could not have served readers who are impatient with the species problem, if I had also tried to present a thorough review of the literature. I am in debt to so many authors who have taught me about genes, categories, and species, but not all of them are mentioned in these pages. I am sorry that I was unable to cite and discuss the work of more authors who have contributed to our understanding of species and SPECIES.

Acknowledgments

This has not been a solitary effort, and I owe a lot to the people who gave their time and insight to help keep me on track. By listening to me and by wading through earlier drafts, and by their critical feedback, they have made this work possible. I am very much indebted to Ernie Lepore, John Maynard Smith, David Hull, John Avise, Jacqueline Stevens, Jerry Coyne, Richard Kliman, Carlos Machado, Jeff Markert, David Houle, Mikkel Schierup, Tracy Solomon, and John Wakeley. I thank Nick Barton and Kathy Roberson especially for lengthy discussions and for thorough, very helpful, critique.

Though it festered for some time, this book actually started to appear on disk and paper while I was a visiting scholar at the Institute of Cell, Animal and Population Biology (ICAPB) at the University of Edinburgh. That sabbatical was made possible by a fellowship from the John Simon Guggenheim Memorial Foundation, the support of Rutgers University, and a National Environmental Research Council Award (U.K.) to Nick Barton at the University of Edinburgh. I am also indebted to the support staff of ICAPB, the Division of Life Sciences at Rutgers University, and the librarians and library staff of both Rutgers and the University of Edinburgh.

CONTENTS

FOREWORD

Do we really need another book about species? I think we need this one. Jody Hey has something new and illuminating to say. He starts by making a point that is obvious once made, but not often appreciated. The nature and identification of species is hotly debated by biologists, and many attempts have been made to find a definition of a "species" that would help to settle the debate. The argument shows little sign of coming to an end. Yet no one treats the question "What is a species?" as one that could be settled by observation or experiment. In other words, they do not behave as one would expect scientists to behave. Why not?

Hey argues that the answer lies in the way human beings think, and, because we use language to think, it lies in the nature of language, particularly in the way each of us must learn the meaning of words. The book therefore does not consist only of an account of the nature of biological diversity, of evolutionary theories about how that diversity arose, and of the many competing definitions of the term SPECIES. It also gives an account both of philosophical and psychological views of language. I cannot summarize his argument in a brief foreword, but I can explain why I found it persuasive. I had already traveled the road that he describes, albeit in a partial and confused manner. The road started when, as a boy of 8, I moved with my family to live in the country. I became fascinated by the birds that were attracted to a bird table in the winter. I soon recognized a dozen or so kinds, but did not know their names. Almost a year later, an aunt gave me a picture book of British Birds. I still remember the excitement of finding pictures of blue tits, great tits, chaffinches, and the rest, and of learning their names. There really were kinds of birds out there, and the kinds I had recognized had names, although I had not spotted that the black bird with a yellow beak and the dark brown bird of the same size and shape were the male and female of the same species. It was the Garden of Eden before I ate the apple of the tree of knowledge.

Some 60 years later, I started collaborating with a group of bacteriologists studying the origin of penicillin resistance in *Neisseria*. I soon learned that a number of species of this bacterium were found in humans, two that cause serious diseases—meningitis and gonorrhoea—and the rest harmless. But as our studies proceeded, it became apparent that genes were being exchanged between all these species, and that the number of species one recognized and the placing of particular isolates in one or other species depended on which gene, or which phenotypic trait, one used in classification. The strains we were looking at simply did not fall into natural kinds. Yet some of my colleagues continued to believe that there must be species, to debate how many there really were, and to which species particular isolates belonged. I was once driven to accuse a friend of being a lumbering robot programmed by his genes to believe in species. Perhaps I was being unfair—I cannot myself avoid using specific names, for example, *Neisseria meningitidis* or *N. mucosa*, when writing about these bacteria.

I think the personal history that I describe will be familiar to most biologists who have studied patterns of biological diversity. Sometimes the organisms fall nicely into natural kinds, sometimes they don't, and sometimes they do until one looks more closely, and then things start to go wrong. Yet even when classifications seem difficult, we feel driven to identify species. Despite my use of the phrase "lumbering robot," I do not think it is really a question of genetic determinism. Jody Hey takes a more considered and more sophisticated view. He was led to wonder whether there is some feature of the way our brains work that leads us to be trapped in the species dilemma and has found that there is indeed such a feature and that it lies in the way we learn to use words.

Hey does not argue that we should stop worrying about the nature of biological diversity and how it evolved. He has spent much of his research career on just such problems, and I hope he will continue to do so. He gives a lucid account of the biological facts and explains why the observed patterns cannot be forced into a single scheme of classification, however ingenious our definition of the term SPECIES may be. The book ends with a discussion of what we should do. I don't think biologists will find it easy to follow Hey's advice because it requires us to recognize that we have a mind-set, arising from the fact that we are a language-using species, that makes it hard for us to think sensibly about biological diversity. But it is better to recognize our limitations than to believe that we can walk on water.

John Maynard Smith

PART I

The Hidden Question

ONE

THE SPECIES PROBLEM

Are dogs one species or many species? Our canine pets and helpers are a diverse lot, but it is not just the diversity that causes this common question. The uncertainty also arises from knowing that so much of dog diversity is found between fairly distinct breeds. When engaged in plain talk, people usually use the word SPECIES to speak of distinct kinds of organisms. In this sense, there are many species of dogs, indeed hundreds of kinds in so far as we refer to different breeds as kinds, and to kinds as species. But we usually stop short of equating dog breeds with species when we recall how easily dogs of different breeds mate with one another (and produce healthy fertile pups that look like a mixture of the parents). Mixed breed dogs remind us of another commonplace view of species, which is that organisms of different species do not join in sexual reproduction. But with this sexual criterion in mind, it is tempting to ask whether breeds of very different size and shape are different species. Does it help to answer the question about dogs and species if we suppose that a Chihuahua cannot jump enough to make puppies with a Mastiff that cannot stoop enough? Probably not. Firstly, reasoning by the mechanics of copulation leads to some puzzling groupings, perhaps including a Chihuahua and a Toy Poodle in one species that does not include a much larger Standard Poodle. Secondly, the most we could ask with this criterion is whether two dogs are in the same species, and the criterion could not be used to count species or to describe species boundaries.

The dog species question is not an easy one to answer, and the difficulty arises in a way that shares aspects with some of the famous philosophers' riddles. The question itself seems to provide little clue of where to seek an answer, and it is easy to go round and round with thought and debate, circumnavigating with multiple distinct attacks on the problem without making much progress. Biologists are also uncertain of how to give names to organisms when diversity does not lie simply.

Consider the variety of opinions that could be anticipated if the question about dogs were put to a series of biologists. Among the possible answers there are four that can be reasonably anticipated, including: "Dogs are one species because they can mate with one another"; "Dogs include many species, as each breed has its own distinct characteristics"; "Dogs represent one of the difficult in-between cases"; and "It depends on what you mean by the word SPECIES." Not one of these answers is very satisfying. As was noted above, the first is wanting on the issue that some breeds of dogs cannot mate with some others because of size and shape issues, and the second is wanting in that it does not help with regard to mixed breed dogs. The third and fourth may be true, but at best they are just the beginnings of answers.

Dogs and their diversity are familiar and thus are useful to help bring on the larger point, which is that biologists regularly suffer great uncertainty over species-related issues. Dogs are a puzzle, but they are but one example of our species uncertainty. The problem could have been introduced using house cats, or oak trees, or blackberries, or Darwin's finches, or viruses, or fungi, or any of the thousands of groups of organisms for which biologists are unsure, or cannot agree on, how to identify species.

Behind the riddle, or *species problem* as biologists call it, are the related uncertainties of how to define the word SPECIES and how to identify actual species in nature. Not for lack of attention does the problem persist, and many books and articles have been written on the subject. Much of the discussion of the species problem has been cast in terms of species concepts, which are competing ideas about what species are—definitions if you will. A recent catalog of concepts found 24 different formalizations that have been proposed and used (some far less than others) over the years (Mayden 1997). A species concept may be fairly abstract and philosophical, even metaphysical, if it is a description of what kind of reality species have. Other concepts are more overtly grounded in biology and evolution. Some concepts contain criteria to aid an empirical inquiry on the presence of species—to help make decisions based on data—while others are more hypothetical and theoretical and seem intended to be useful in discussion but not in application. Among the former, practically oriented concepts, some have been honed with care to be as terse and unambiguous as possible. Unfortunately, there persist several such carefully crafted concepts, and it is not clear which, if any, should be used. Some authors have even proposed that the best species concept is a polymorphic one that supposes species to be different things in different contexts. Even the most persistent of polite and careful debates end without resolution, with each proponent persisting in the claim that their own concept is best (see, e.g., Wheeler and Meier 2000). Far from a shortage of species answers, we have a wealth of verbiage, meanings, and ideas on which to draw when we inquire of biological diversity. We have lots of ideas and answers, but most noticeable is our persistent and awkward shortage of consensus.

My own research on the origins of species suffers from these uncertainties. With students and other scientists, I have investigated the genetic differences and similarities of closely related species of fruit flies. Our main goal has been to elucidate the different ways that evolutionary forces have given rise to multiple species. We have pursued this by examining the genetic similarities and differences of samples of flies of so-called different species. They are "so-called" because it was other biologists who have given species names to flies that live in particular places and that share certain characteristics and that we have studied. Our pursuit began by drawing on that prior research and by using those previously assigned species names and the observations that inspired them. We did find that often there are considerable genetic differences between flies of different species, but this is not always the case. Once we found essentially no differences between samples of different species—no more than expected by chance between two different samples of one species. Another time, we found considerable differentiation among strains of what had been identified as just a single species, suggesting that it actually included more than one species. Other times, we found complex patterns of variation that did not lead to simple interpretations of species status. The immediate impression of these uncertainties is that previous species designations are often not borne out in patterns of genetic differentiation. By itself, this finding is not a surprise (older species designations often do not bear up under closer scrutiny), and it is not necessarily troubling. There is no particular reason that species should be easy to identify, and perhaps an entirely resolved species problem would still leave us with words and concepts that present challenges in application. What I found to be the greater difficulty raised by my research is that it has not lead to a greater understanding of how to identify species, or of what SPECIES means or should mean. Even as we have gained insights on the evolutionary forces associated with diversification, I am not at all assured of how species should be identified or of what the most suitable definition of SPECIES might be. Many other species inquirers have tread similar paths and learned a similar frustrating lesson.

For biologists like myself who suffer it, the species problem arises mostly in one of two contexts: We say SPECIES, but cannot easily explain a meaning; or we articulate a meaning for SPECIES, but then find that other biologists do not agree. When presenting the results of my research, I use SPECIES, but if I am asked what I mean by the word, then the only short answer I have is to say that we are working with the labels—the species names and descriptions—that others have applied and that have been generally agreed on in the past. At times I have been accosted for using SPECIES in a way that some did not find consistent with my data. Depending on the details of the data and the accuser's particular definition of SPECIES, I often have no defense but to say that the accuser's definition of species may not be ideal—a weak defense at best, if I am not able to promote a good alternative.

The History of the Problem

The words SPECIES and GENUS, which we use to help us refer to kinds of organisms, come to us from their use in classical logic as laid out by Aristotle. They are translations of the Greek words EIDOS (εἶδος) and GENOS (γένος) that were used by Aristotle and Plato to refer to categories, or kinds, of things. For the general purpose of metaphysics, of questions about the true nature of things and of their existence, a genus was a category, and depending on its contents it might be differentiated into more narrow categories called species. The contents of a category were also not necessarily smaller categories, but could also be individuals (meaning real entities, and not necessarily a person or an organism). From the time of the Greek philosophers up until the 18th-century Europeans, SPECIES was also used outside of formal logic, with a variety of meanings having to do with the appearance of something. This dual usage actually dates at least to Aristotle and Plato, who used SPECIES not only in a formal philosophical sense, as a class of things, but also with a much less formal usage to refer to the form of things (Ross 1964). Similarly, genus was used to refer to a level of category—more inclusive than species—but it was also used less formally to refer to a race of organisms.

Since the time of the Greek philosophers, GENUS and SPECIES have seen a great deal of use in application to kinds of organisms, but these meanings have not always been very precise. One reason for this was the commonplace view that kinds of organisms were not distinct. Prior to the 18th century, ideas that organisms arose by spontaneous generation and that some organisms could give rise to others of different kinds, were commonplace (Zirkle 1959). Our current knowledge that all organisms come from very similar parental organisms did not gain broad acceptance until the generality of "true-breeding" and the falseness of spontaneous generation began to become known. With the flowering of biology in the 18th century, stricter biological meanings for SPECIES and GENUS began to take hold (Mayr 1982). It was Carl Linnaeus, especially, who solidified the biological meanings of these words by describing kinds of organisms in the terms of a strictly imposed formal system borrowed from classical logic. We can thank Linnaeus for a three-pronged attack that transformed biology into a field with strictly codified methods of reference to kinds of organisms. In the first place, he held that genera were distinct and unchangeable kinds of organisms that had been created by God. (He also said the same of individual species, but his views on the fixity of species were less certain later in his life.) Thus each genus did represent a true kind of organism, with a God-given essence. Second, Linnaeus gave genus and species names to virtually all plants and animals within his experience, which was very broad, as students and collectors regularly shipped samples to him in Sweden. Finally, he described practical methods for identification, by which individual organisms could be correctly named as members of particular previously described species. His sexual system for identifying plants, which he developed early in his career, has proven to be one of the handiest tools in the history of biological classification. The conventions of using GENUS and SPECIES that Linnaeus described, as well as many of the individual names of genera and species and the preferred methods for the identification, survive to this day. What also survived up until, but not past, the time of Darwin, was the view that organismal kinds are fixed and unchangeable.

But despite the heroic efforts of Linnaeus, there was in Darwin's day a version of the species problem. It was manifest primarily as a lack of agreement among naturalists on how to identify, distinguish, and count species. If Darwin were still with us, he would probably be vexed that the modern-day species problem is a thriving direct descendant of the one he faced down in 1859. Darwin thought that he had explained what species are, and that he had also explained why the identification of species is difficult and inherently partly arbitrary. But despite his fame for the discovery and explanation of natural selection, his views on species have mostly gone unappreciated. In fact, Darwin has earned a bit of notoriety for a particular irony that has developed since the publication of his book. On the one hand, he succeeded in convincing others of the historical fact of evolution and on natural selection as the mechanism of evolution (though it took decades, and Darwin was long dead, before biologists were in agreement on the topic of natural selection). But on the other hand, he failed, for he was not generally persuasive on the actual nature of species. (Hull 1965a; Sinser 1950, 305; Wallace 1901, 152.) Darwin did not see species as a special kind of collection of organisms, distinct from varieties or subspecies. To him, SPECIES was just a word that is applied to a grade of distinction, not unambiguously identifiable in many situations, and certainly not different in theory from other grades of distinctions. "Many years ago, when comparing, and seeing others compare, the birds from the separate islands of the Galapagos Archipelago, both one with another, and with those from the American mainland, I was much struck how entirely vague and arbitrary is the distinction between species and varieties" (Darwin [1859] 1964, 48).

Darwin goes on to describe how the continuum of variation—individual differences, to varieties, to species—is not one in which naturalists have been able to make agreeable distinctions. He saw that repeatedly consensus fails on the details of how varieties and species are to be delineated. Darwin interpreted this persistent lack of agreement on species boundaries as evidence that species have a history that includes the formation of varieties as arising by the same process that gives rise to species. The disagreements among naturalists, about the way organisms should be grouped, were a clue to Darwin that it is the variation among individual organisms, especially those variations among similar organisms that are identified as being of the same species, that is the ultimate source of all of life's diversity. Over time, some of this interorganismal variation becomes manifest as differences between varieties (what others have called subspecies, or races) and then with still more time, these differences become such that the label of SPECIES becomes useful. But Darwin envisioned no quality to individual species or to the species category that should lead us to perceive it as special or different in kind from other taxonomic categories like genus or subspecies.

The irony is that many modern biologists disagree with Darwin. They hold that species are distinct real things in nature, different from varieties, and with a certain kind of history and with identifiable boundaries. This modern view should not be confused with the view that

Darwin successfully attacked, that each species represents a nonevolving essential type, distinct from other types. To Linnaeus, who articulated this essentialist view, each type represented an instance of creation. Present-day biologists firmly reject essentialism, but many do think of species as largely distinct and as having special biological properties— and with these views they differ from Darwin.

The changing views on the distinctiveness of kinds of organisms can be roughly grouped into four stages. (1) Prior to Linnaeus, GENUS and SPECIES did not have distinct meanings in reference to organisms, and organisms were not always thought of as being members of distinct and unchanging kinds. (2) Linnaeus changed all that by formalizing GENUS and SPECIES and by insisting that kinds of organisms are distinct and unchanging entities created by God. (3) But this static and comfortable view did not resolve many cases of uncertainty. In laying bare the causes of uncertainty, Darwin's book lead to a revolutionary overturning of the entire edifice. In short, Darwin said that there are not distinct kinds of organisms for the simple reason that organisms evolve. (4) In the time since Darwin, biologists have embraced the idea that organisms evolve, but have returned to the idea that organisms occur as members of distinct kinds, though with little agreement on the properties of those kinds. Darwin was triumphantly persuasive on evolution, but not on his view that species are not distinct. The view that organisms do occur as part of distinct kinds has been held by some biologists ever since Darwin for various reasons and under various models of distinction. It is probably the prevailing view today, but it is hard to tell because of the disagreements on the meaning of SPECIES.

From this brief history we can partly see where our peculiar dilemma, the common use of a word with no common meaning, has come from. The common meaning we once had has been knocked aside by our modern understanding of evolution. Darwin has left us stranded on a word, and modern biologists are semantic castaways, trapped with a word of little common meaning, struggling to fix the situation by puzzling their way out of it.

Living with the Species Problem

The species problem is famously difficult, and some biologists have despaired, concluding either that there is not an answer, or not one answer, or that the answer is inaccessible. Thus Levin (1979) was inspired by the difficulties of trying to make sense of plant diversity to propose that species are not more than tools, abstract constructs that help us refer to diversity. Endler (1989) partly embraced this view and elaborated on several different categories of meanings that were associated with different variants of species debates. Endler claimed that we are not all trying to explain the same thing, and that a part of the species problem is a confusion that we are.

Besides Levin and Endler, numerous authors have proposed that the species problem is not simply resolvable, and that we should recognize this and move on. There are two central bases for such proposals. First is that the large species problem may be insoluble in a conventional way. For biological reasons and because of the lack of progress on the problem, it certainly seems unlikely that one methodology and one definition of SPECIES could ever be agreed on. Consider too the truly staggering amount of diversity (e.g., biochemical, morphological, behavioral, and genetic diversity) that occurs among living organisms. Perhaps it is simply unreasonable, and unrealistic, to even suppose that one word, with one definition, could be used across all the breadth of that diversity. The second reason for the apparent sensibility of proposals that we drop the quest for a single all-encompassing solution to the species problem is the evident progress of biological research, even in those areas of biological diversity that might be expected to suffer most for species uncertainty. The pace of progress is impressive in various ways, and biologists are clearly not simply enfeebled by the problem. They have a working-definition method of progress, in which each researcher uses a species concept that seems reasonable for the organisms being studied at the moment. They may not even articulate this idea in their writings, or even to themselves, but they have a sense of what species are, for the organisms they study. Thus biologists who study the formation of species in animals will probably share a view of species, whereas those who study plants that cross hybridize more freely will have a different meaning, and those who study things that rarely, if ever, exchange genes will also have a distinct view, and species-level systematists will have yet a different view. Actually, these example categories are too large, and even within these specialized areas there is only limited consensus. But on a day-to-day basis, biologists interact mostly within small communities that share interests and outlooks, and the bulk of the dissemination of results goes on within circles at scientific meetings and in journals of modest scope.

The method of multiple workaday solutions to the species problem has also been codified, though with only limited success. Scattered throughout the biological literature are reports of the existence of species *sensu* the concept of the moment. There are a fairly large number of summaries of the form "The meaning of SPECIES for the *X*," where "*X*" is replaced by the name of some large group of organisms. So too have a number of very reasonable words been proposed for specific contexts, such as: agamospecies and paleospecies (Cain 1954); genospecies, nomenspecies, and taxospecies (Ravin 1963); and cytospecies in parasitology (see, e.g., Lane 1997). For plants, a need was foreseen for a field-biology terminology that allowed reference to closely related groups of organisms that did not run afoul of, or strongly depend on, the uncertainties of species identification. Gilmour and colleagues devised the -DEME terminology for reference to various groups of closely related plants (e.g., gamodeme, topodeme, ecodeme) (Gilmour and Gregor 1939; Gilmour and Heslop-Harrison 1954). These -DEME words and the other -SPECIES words were created with clear meanings, and they are sometimes used to considerable effect, vastly clarifying a discussion that would be more muddled if only SPECIES were used.

But despite the apparent success and utility of these practical resolutions, the species problem still flares up; particularly when biologists of different stripes interact and when questions arise that expose the uncertainties of the working-definition of the moment. The working-definition approach is also far less useful for biologists who work simultaneously on multiple very different sorts of organisms. What underlies these persistent conflicts is a lingering quest for a common method and meaning.

It has been my experience—and I am guessing that it is a typical one—that when talking with biologists, one hears SPECIES tossed about regularly in a manner that supposes there is one single common meaning. If pressed on that common meaning, biologists are stuck, but they persist in using the word in a casual way much as laypersons do, as if it has a well-known meaning. Other evidence that biologists still seek one meaning comes from the many articles and books that begin by posing the question "What are species?" with an implication that a single meaning is a desirable property of the answer. Consider also the limited appeal of various proposals for breaking up the different meanings of species that have been proposed and giving them different names. The various narrow descriptions of species, the codified working definitions, have not been so successful that biologists generally recognize that this approach carries the seeds of a long-term resolution. Gilmour's -DEME terminology came to be used infrequently, and then in a fashion that was incorrect according to the proposer's guidelines (Briggs and Block 1981). Briggs and Walters had used the -DEME terminology in the first edition of their widely used reference on plant variation (Briggs and Walters 1969), but because the method was not accepted by biologists, it was dropped in the 2nd and 3rd editions of the book (Briggs and Walters 1984, 1997).

At the heart of the failure of the -DEME terminology and of the infrequent use of other narrow meanings of SPECIES is a curious behavior in which biologists who discuss the meaning of SPECIES all accede to somehow try and talk about the same thing. When scientist A says species are X, and scientist B says species are Y, typically neither A nor B would be happy with the answer that some species are X and some are Y. Species problem solvers usually propose just one definition, and debates about species are cast as a competition among alternatives. These alternatives are mutually exclusive only in the sense that there is a search for a best concept. They are not mutually distinct, such that only one can be entirely correct, and all others must be entirely incorrect. There is a partially hidden consensus that we are all trying to explain the same thing and that we can only partly see that thing. Hardly ever does the proponent of one concept claim that another is simply wrong. Rather, the typical claim is that other concepts, while on the mark in some respects, are not complete or ideal under some criteria. It is as if we are all trying to figure out the true nature of the strange, partly invisible thing on the table. Everyone trying to describe it is convinced it is there, and quite reasonably nobody is satisfied with simply calling it a strange, partly invisible thing. The structure of the debate is not really this tidy, but it is strange in that so much discussion concerns the merits of alternatives that are all pointing in the same direction.

In a sense, biologists are still holding out for the Big One, that one single definition of SPECIES, or that single succinct species theory, that will dispel our questions and squabbles. Some of this pigheadedness is reasonable under the conventional scientific method, for that method favors a search for uncomplicated answers first. Thus, conclusions like Endler's, that we cannot find a single answer to the species problem, still beg the question of whether our lack of progress on answering the question is simply an example of what happens on the way to solving a hard problem, or whether our lack of consensus is somehow necessary and inevitable because there cannot be one answer. The claim that it is unproductive to try and answer the big question with one answer is a statement about history. It has not been productive so far, but that does not necessarily mean an answer is impossible. Problem solvers of all sorts must pursue simple explanations first and more-complicated solutions only after the simple ones have been rejected. As much as one may realize the possibility that no single meaning of SPECIES could do justice to all the different purposes that have been envisioned for it, as long as we do not know that such a concept is impossible, the fruits of such a single meaning are very tantalizing.

Are We Missing Something?

The scientific endeavor is fundamentally one of answering questions about the natural world. Many such questions loom large in the minds and communities of scientists, and some have rich and lengthy histories, as does the species problem. But the species problem is not like other scientific conundrums. It has an unscientific strangeness that sets it apart from other questions. The debate over species is unusual in what it lacks. It is very much a debate among scientists, and yet no part of the debate concerns that most basic question at the heart of nearly all scientific inquiry. Species debaters do not ask, "What information do we need to help solve the problem?" This question is fundamental to all problem solving, and it is not just for scientists. Consider any problem or challenge that you must solve and to which you are lacking a solution. Do you not ask, "What am I missing?" or "What do I need?" or something like that, as a way to get started? This general casting for knowledge is nearly ubiquitous in all manner of inquiry, and if the question is not posed explicitly, it at least seems to be a reasonable question. But such is not the case for the species problem, where virtually none of the debate touches on our lack of information or the quality of our current knowledge base. We think we have an entirely adequate understanding of evolution that is both necessary and sufficient for understanding how biological diversity may arise. In brief, it seems clear from the mass of theory and data on biological diversity that there are five major factors involved in the evolutionary rise of biological diversity among organisms. The factors are mutation, natural selection, genetic drift, geographic separation, and genetic recombination. It is true that in fairly few cases have the relative roles and detailed portraits of these factors been elucidated, but invariably some combination of these factors seems sufficient to explain observations.

Yet amid this broad general knowledge, biologists are still stumped on species, so firmly stumped that most are not even asking how to get unstumped. We have researched and written and debated ourselves as far as we can go—or so it seems—and that destination is a dead end. There is no literal obstruction, no impenetrable reason not to ask of what information we lack. Rather, we seem to have run out of ways to look for gaps in our understanding. We have reams of data and shelves of books and journals on all manner and manifestation of biological diversity. Biologists have also chewed on dozens of definitions of SPECIES from seemingly every which way. So much has been done, and so much has been learned, and yet the species problem stands; and it stands pretty much as firmly as it did in Darwin's day. The debates continue, but they have a disheartening repetitive aspect (e.g., Wheeler and Meier 2000) that comes from working without new information and without progress. We work as if the species problem were one of those hand puzzles in which two or more intercalated parts must somehow be untangled from one another, and the only method is to fiddle about with what we already have before us.

It is not that species debaters have given up. Even though some may refuse to engage the issue, there are many others of us still looking and hoping for the Big One—still searching for that magic bullet that will make our headache go away. But the forms of the debates have been cast for so long that it is hard to see beyond them and to ask again of what information we need.

This book is the outcome of asking, "What is it that we require if we are to resolve the species problem?"

TWO

THE MODE OF IGNORANCE

In the previous chapter, I urged that we not suppose that we already know everything that we need to know to address the species problem. I said that if we are to get unstuck, we must look for something that is missing from our inquiries and from our knowledge base. But some would say that this is wrong, and that rather than continue to seek a solution to the species problem we must recognize that we are quite properly stuck and must remain so. This is a patently reasonable position, and some very reasonable biologists, like Levin and Endler, have said as much. The primary pieces of evidence in support of this contention are all those reasons why we are stuck. In many ways, we know why species are hard to identify and why SPECIES is hard to define. Our failures have taught as a great deal about how hard the problem is, even if they have not taught us how to solve it. Secondary evidence, in support of an "unsolvable" stance on the problem, is that our workaround for proceeding without a solution, the method of working definitions, is quite a productive one. It is far from true that no communication occurs among biologists, and the working-definition method somehow enables a great deal of research on biological diversity. Perhaps we are making the most of an insoluble conundrum. If so, then we should appreciate the species problem as another unpleasant fact of life, rather than as something that needs to be fixed.

This chapter has two purposes. The first is to address, and reject, this argument that we have already achieved much the best resolution that can be achieved. As reasonable as this idea may be, it is wrong; and the reason that it is wrong is central to the process of measurement and to the method of science. The rejection of the "unsolvable" viewpoint, and the affirmation that the species problem is a conventional scientific puzzle, will also give us a way to proceed. It will give us a way to ask the question, "What information do we need?" and to begin a new search for information that will help address the species problem. That is the second purpose of this chapter, to start a new investigation of the problem.

Measurement and the Scientific Method

In the course of coming to understand the world around us—in general and not just regarding species—we observe and ponder, and we measure. This version of MEASURE goes best with a predicate, and measurements are made of stuff or force or things; always some feature of the world that we have cogitated into measureableness. Potential angels on pinheads do not meet this standard, and those things that we call subjective can bear only the crudest measuring process (which is why we call them subjective). Scientists struggle to find measureableness, to learn of features of the world and ways to measure them effectively. But why do they undertake measurements? What is it about measurement, that they see the process as crucial to their endeavors? To answer this, consider that making a measurement amounts to forming some kind of connection between an idea in the mind of the measurer and the world outside that mind.

When we think about the process of measurement, and of the purpose that measurements serve, it is useful to recall a central point about language that has been made by a number of philosophers, which is that a person's words necessarily have two different kinds of meanings. The personally accessible and upfront content of the thoughts, the meaning that a person tries to convey with their words, is what has been called the connotation, or the intension, or the sense of the words. But language about the world, and thoughts about the world as well, have a second, more basic meaning which is that aspect of the world that is pointed to by the first. This more external component of meaning has been called the denotation, or the extension, or the reference. The reference is that feature of the real world that caused the sense; it is the reality that the sense, and the language describing the sense, are about. A body and a mind have built the sense, and it need not be a very good description of the reference. Probably the most famous example of an inaccurate sense, from olden days, was of a ball of fire in the sky and that was commonly associated with the word SUN (or its counter part in another language). Of course, examples are innumerable, for they arise on a daily basis in hindsight whenever we see the discrepancy between an idea we have, and the reality of what we thought we referred to. To be strict, we do not ever literally know the true reference, as we can never know reality apart from our perception of it, but the point still stands if we consider how our sense of things changes as we gain more information and understanding.

An apt summary from Deacon (1997, 61) is that sense is something in the head and reference is something in the world. Clearly in this light, the mission of anyone who would honestly try to describe reality is to bring their sense to the reference, to have them not be contradictory. This is where measurement comes in, for it is a bridge built from the sense to the reference. The bridge builder may initially have little insight to the quality of the measurement, but future measurements built from the same sense by the same builder can only be fundamentally similar to the first if the sense-reference connection does not change; so, too, must the measurements by a different person, who has a similar sense of what is being measured, be commensurable with those of the first if the sense is to be maintained. Repeated measurement is not simply a tool for checking one's sense of magnitude; it is also a check on whether that sense is not an entirely aberrant impression. Depending on the context and the outcome, measurement may affirm, or adjust, or overturn the sense that inspired it.

This description of measurement is but one way to describe the empirical foundation of our knowledge of the world. Although I have described measurement by way of bringing out the distinction between sense and reference, the core idea is neither fancy nor something that is limited to scientific research; it is simply that we cannot learn more about the material world without somehow checking the correctness of our ideas about it. This is just as true for an idea about the mass of neutrinos as it is if we wonder how much beer is in the refrigerator. Whenever measurement is undertaken, there is a hypothesis being tested (Popper 1959), though it may be small and implicit or large and explicitly acknowledged. At base, we cannot understand the world, or trust our sense of it, if we cannot make measurements that confirm that sense.

Now we have at least one way to evaluate the cost of the species problem. To be plain, if understanding is good and ignorance is bad, then we can judge the state of affairs by the consistency of our measurements. Furthermore, to the extent that we find inconsistent measurements, it is fair to conclude that our understanding can be improved and that we suffer for the lack of it. To say otherwise is to claim at least one of two profoundly unscientific things: either that inconsistent measurements are not a sign that understanding falls short; or that reality varies with the observer. For those who despair of resolving the species problem, and are aware of the many ways that measurements regarding species are not consistent with one another, the argument that is laid here should come as good news. The very fact that we can make measurements, and that we can see them to be inconsistent, tells us that we can improve our understanding that leads to those faulty measurements—there is necessarily a reality that is more consistent than our current understanding of it.

Measurement Problems

Of course not all measurements of species are inconsistent, far from it as evidenced by the considerable success of the method of working definitions. But biologists who make measurements of species diversity or who rely on measurements of species diversity regularly run afoul of three fairly conspicuous arenas of inconsistency, three recurrent kinds of measurement problem that are common enough to give names to. I label them COUNT CREEP, CONCEPT CONFLICT, and FUZZY SPECIES. Count creep, is what often happens when a biologist gathers new data, and takes a closer look at one or more previously recognized species. Very frequently they find evidence for a new pattern of variation within what had previously been recognized, and thus evidence for more species than had previously been recognized. The inconsistent measurement is simply the species count, and it is steadily being revised upward the more that biologists study diversity. Recently, for example, the species of small fruit fly called *Drosophila melanogaster*, arguably the best studied species on the planet (apart from our own, *Homo sapiens*), was found to include a second species that lives in parts of Africa (Hollocher et al. 1997a, 1997b; Wu et al. 1995).

Evolutionary biologists are used to the business of creeping species counts, for it happens quite frequently. One common view, rarely articulated but clearly implicit in a great deal of research, is that it is a process that must stop as larger and larger samples are used in the measurement process. But such a view is incomplete. In the first place, there is no certain way of knowing of unrevealed species within those already recognized, no certain way of knowing that the creeping will stop, unless all of the members of a species can be thoroughly studied. Secondly, it does not explain the typical certainty and definiteness that is associated with many species descriptions, names and counts. Most summaries of the kinds of organisms that exist, in some place or habitat, provide these descriptions without any acknowledgment that the species descriptions and counts are likely to be overturned with larger studies and finer assessments—just as previous studies have been overturned.

The second kind of measurement inconsistency arises from concept conflict. This occurs when one biologist makes measurements on a sample of organisms, while thinking that species are one thing, only to have another biologist come along and record different values for the same organisms (often the very same individual organisms), while thinking that species are something else. The most general concept conflict occurs between "lumpers" and "splitters," between those who somehow feel that small differences among organisms do not necessarily make species and those who feel that differences are the essence of species' distinctions, even if they are small. Concept conflict is not new, and it was part of Darwin's starting point in *On the Origin of Species*. Darwin began his argument for the evolution of species by drawing upon the failure of naturalists to know how to, or agree how to, distinguish and enumerate species: "we find that there are hardly any domestic races, either amongst animals or plants, which have not been ranked by some competent judges as mere varieties, and by other competent judges as the descendants of aboriginally distinct species" ([1859] 1964, 16).

And he brought the point home rather forcefully, with the help of leading botanists:

> Compare the several floras of Great Britain, of France or of the United States, drawn up by different botanists, and see what a surprising number of forms have been ranked by one botanist as good species, and by another as mere varieties. Mr. H. C. Watson, to whom I lie under deep obligation for assistance of all kinds, has marked for me 182 British plants, which are generally considered as varieties, but which have all been ranked by botanists as species; and in making this list he has omitted many trifling varieties, but which nevertheless have been ranked by some botanists as species, and he has entirely omitted several highly polymorphic genera. Under genera, including the most polymorphic forms, Mr. Babington gives 251 species, whereas Mr. Bentham gives only 112, a difference of 139 doubtful forms!" (Darwin [1859] 1964, 48)

If we turn to the modern field of species level systematics, wherein biologists work at enumerating, classifying and naming species, we see that the concept conflict still persists. Systematists are steadily in the business of revising the classifications of earlier systematists, and the field is still full of lumpers, who revise species counts downward by lumping multiple species, and splitters, who revise species counts upward by splitting single species into multiples. These debates are over how to partition a single set of observations, and so they are different from the species splitting that happens with new data (that is count creep). Consider the case of *Metriaclima*, a genus of 10 species of fish in Lake Malawi, Africa, that was devised to replace a single species, *Pseudotropheus zebra*, on the basis of additional collections (Stauffer et al. 1997). This first revision was a case of count creep, pure and simple—a closer look with more samples begat more species. But then another look, by others who were using the very same data as Stauffer et al., lead to the conclusion that the new genus actually contains only two species (Konings and Geerts 1999). Nor are lumper/splitter debates limited to obscure organisms that are difficult to collect. Consider birds, which are probably the most observationally accessible animals on the planet (from peoples' perspectives). Conventional classifications place the number of bird species worldwide at around 9000. But some ornithologists feel that the correct count, based on a proper reevaluation of all existing collections, would end up being closer to 20,000 (Martin 1996; Zink 1996). In fact, a count of endemic Mexican bird species went so far as to employ two different definitions of SPECIES; one returned a count of 101 species while the other returned a count of 249 species (Peterson and Navarro-Siguenza 1999).

The third arena of inconsistency is that of fuzzy species. Just as with fuzzy sets and fuzzy systems, a fuzzy species is one that is indistinct and without sharp boundaries separating it from other species. Fuzzy species are common, and many species are very fuzzy. Probably the most well-known and well-puzzled-over examples are the many plants for which species designations often seem arbitrary. Plant species that seem distinct by some criteria very often include individuals that have exchanged genes fairly freely with plants from other species. Higher plants are famously problematic this way (Grant 1981; Stebbins 1950), but in recent years many similar cases have been found in other groups of organisms. Indeed so many species have vague boundaries, and are only partly distinct from other species, that this general uncertainty probably makes up the largest arena of inconsistent measurements. Suppose, for example, that the other two problem areas, creeping counts and concept conflicts, could be resolved in the future. Perhaps with lots of data and lots of discussion, these issues would gradually die away, and we would have consistent species counts. But any hopes for such a world are dashed when we learn about how many species are fuzzy, and about how fuzzy they are.

There are several reasons for recognizing that real species in nature often do not have sharp boundaries that separate them from other species. First, Darwin's explanation of species was cast as an explanation for why the species category is not useful; i.e., species are not different from varieties because they emerge gradually from varieties. This basic idea, that most species emerge gradually, simply cannot be ignored, no matter how troubling we might find its most basic corollary—that species are often not distinctly separate from closely related species. Second, the best theories of species formation, that have emerged since Darwin, also lead necessarily to the idea that there will sometimes be in-between cases where it is not appropriate to say either that there is one species, or that there are two species. Third, and finally, a number of the best studied species and cases of species formation have revealed many examples of just the sort of in-between circumstances that are expected on the basis of these theories.

The difficulty with fuzzy species is that they cannot be counted, at least not with any consistency of measurement. Until now, in this book, it has been convenient to mostly adopt a SPECIES parlance in which species are assumed to be entities with real distinct boundaries, and thus that they are countable. But fuzzy species do not have distinct boundaries. The counting of the number of species is simply, and frequently, the wrong way to quantify biological diversity.

What is most strange about fuzzy species is the way they are ignored. Darwin's conclusions, and those of many other biologists, and the evidence from many studies of indistinct species, are not news. Most biologists *know* that species are often inherently uncountable. Indeed, a major undercurrent of species debates is a recognition (widespread, though it varies in degree) that biological diversity is not always patterned into distinct, countable species. Stebbins (1969) was so taken by biologists' insistence that nature have objective distinct species, in the face of ready evidence that it does not, that he devised an allegory—a fairy tale to draw attention to biologists' false assumptions of distinction in nature and of their own objectivity. For some reason, the common knowledge that nature is not made up of objective, distinct species has not prevented biologists from using species counts as their main index of biological diversity. There is a kind of built-in inconsistency, on the one hand we know and acknowledge that species are often not countable entities, and yet we use SPECIES just as if it refers to countable things.

Here is another way to appreciate the issues of fuzziness and counting. Let us look beyond our current entanglements and suppose that in the future we will somehow see the species category more clearly and understand species more thoroughly in some complex entirety. How then might we refer to them? Suppose, for example, that this understanding includes the insight that the species category contains distinct biological entities of a certain kind, and that we learn that some organisms actually constitute species in this sense. And suppose also that we come to understand that there are many times that no such distinction can be applied or inferred from patterns found in nature, that thorough investigations do not reveal the presence of identifiable distinct entities that might be called species in an individualistic sense. I am asking readers to envision what it would be like if we could leap ahead to a time of an improved understanding of species, and I am asking them to suppose that our revised vision is like the two-faced view that I have described—sometimes organisms constitute real species that are distinct and sometimes they do not. With this supposition in mind, we can fairly easily take the imaginative leap to a future world that is comfortable with that kind of knowledge. The reason that this exercise, this leap of thought, is not difficult, is that we presently have a good understanding of some other things that are like species in this two-faced sense. For example, consider our present understanding of clouds— yes, the damp, atmospheric things. Meteorologists have a pretty good understanding of what clouds are and of how they arise, and even a biologist can see clouds both from a distance and from within, when they occur as a surrounding fog. We know they are made of zillions of suspended droplets of water that have condensed from the surrounding air. We also see that sometimes clouds occur with fairly distinct boundaries, and we can count them. At times a cloud is an entity in the sense that it has a distinct location and history. But just as often we see massive cloudbanks or wispy cloud tendrils without any kind of distinct boundaries between individual clouds. Indeed under such circumstances, we tend to use CLOUDS not to refer to some multiplicity of individual things, but rather to an abundance of cloud stuff. CLOUDS is widely used, but no person familiar with the sky and with fog would suppose that it is always used to refer to entities with distinct boundaries and histories. The counting of clouds is a game for some fine weather days in which circumstances happen to give rise to distinct clouds. But imagine the errant meteorologist who spoke strictly in terms of countable entities. Maybe she would say something like "Clouds are moving over our region today. We think there are 42 of them. However, most are so close together that we are not sure of how many there really are. We will keep you posted!" This is silly, of course, but the point is that it seems silly precisely because we have a good understanding of what clouds are made of and that they are not necessarily individual things.

Now return to the problem of species, for which our understanding is woefully behind that for clouds. If it turns out to be true that species are sometimes real, but that they do not always occur as distinct things for many organisms, and if such an understanding became widespread, then any technical discussion that relied on species counts would seem as quaint as our ignorant meteorologist. Or would it?

The punchline of this exercise is that we are already in this situation to some extent. The basic speciation theories that predict indistinct species, and the common evidence that species are not always distinct, tend to be part of the basic evolutionary biology graduate school curriculum. Every evolutionary biologist knows this stuff, even though they vary in their view of the prevalence of indistinct species. Also, do not forget that biologists are familiar with the lumper/splitter debates and with the many observations of creeping species counts. But even though we expect species to often be indistinct, and even though we often find them to be indistinct, and even though we are terribly inconsistent at identifying and counting species, biologists persist in counting species. What is going on here? For some reason, despite evidence to the contrary, "species are countable" is the default position for most biologists.

A New Search

The problems of measuring species may be a steady frustration for working biologists, but at least we can recognize some common features of those problems. So too can we learn from those common features and try to turn them to our advantage. In the first place, this review of the process of measurement, and of the kinds of measurement failure, reminds us that some sort of resolution of the species problem must be attainable. Our reality checks—our measurements—may be failing, but as scientists we must recognize that the failure is ours and not reality's. Somehow there is a way to fix the species problem, and in so doing cause our future measurements to be consistent with each other. But how? Perhaps the sweetest benefit of our many failed measurements is that they do give a clue about where to turn in our search for a new approach.

In reviewing and writing about the measurement problems, I noticed three common threads that seem to pervade the measurement failures. The first two are just what we usually think of when we describe the species problem: our uncertainty about identifying species; and our uncertainty about defining SPECIES. We are familiar with these issues, and that familiarity is precisely why we doubt that additional data on diversity, or additional cogitation on definitions, is likely to lead to new insights. It is the third thread of commonality that suggests a different train of thought. The third common aspect to the difficulties of count creep, concept conflict, and fuzzy species is biologists' knowledge of those problems. Biologists who fail in their measurements, because of these issues, are already well trained and quite knowledgeable about these issues. Biologists who study species diversity know full well that additional species are regularly found within those already known; they know all about lumpers and splitters and that many other biologists will not agree with their counts or their methods; and they certainly know well that many species are not distinct. These difficulties are not new, and still biologists forge ahead, committed to counting species and committed to repeating the same sorts of errors they have suffered in the past. This determined wrong-headedness is a clue about where we should search for additional insights on the species problem.

To be plain, I wonder if there is something about the way that our brains work that hinders our investigation of biological diversity and prevents us from finding our way out of the species dilemma. As awkward as such a notion must seem to many, accustomed as we are to taking our objectivity for granted, it is an idea that is firmly motivated by the peculiar nature of the species problem, and it is an hypothesis. It is also an idea that is suggested by, or is implicit in, the writings of a number of authors (Endler 1989; Hull 1997; Levin 1979; Stebbins 1969). We can still be scientists when examining this idea, just as we would for any other. From here on out, this book will consider not only the nature of real species but also the ways that our minds work to describe them.

But how, exactly, to proceed? The idea under examination, that we carry a mental bias that misleads our investigations of biological diversity, is vague. Perhaps worse is that at the center of the hypothesis is the human mind, our ideas of which also carry a variety of uncertainties. A vague hypothesis and a fairly inaccessible test subject mean that we must proceed with care. Fortunately, much work has been done in other fields of study that bears on the issue of how the mind considers biological diversity. The fulcrum for many of these investigations, including some from philosophy and psychology and anthropology, is not the mind so much, but language. As close as language and the mind must be to one another, our communal behavior of language is wonderfully accessible to investigation, whereas the workings of the mind are not so plain. Necessarily, then, there are many situations in which investigators with a basic interest in the way the brain works use the substance of language as a proxy, with a kind of working assumption that speech and writings, while not the same as thoughts, are closely tied to them.

Words About Words About Species

An important inspiration for the twin themes (species and our language about them) lies in a famous book entitled *Word & Object*, by Willard V. O. Quine (1960). The considerable esteem held of that work by philosophers flows from the progress it makes in disentangling some sensible explanations and insights about language, using language. It is a bootstrapping *tour de force*. Quine proposed that we cannot approach language simply by dissecting words and meanings, that to do so presumes that language is a good tool for dissecting words and meanings—in short, that it has meaning. The assumption of content in language is not one that can be directly attacked. Rather we can explore language, with language, only by going slowly and maintaining consistency. This route does not deny that we may ultimately overturn conventional ideas about language, but "we may kick away our ladder only after we have climbed it" (Quine 1960, 4). Quine's method began with and maintained an essentially naturalistic approach to the philosophy of language, that a philosopher of language is as much in the world as is his subject. If we are to understand our subjects of study, then we should also be prepared "to ponder our talk of physical phenomena as a physical phenomenon, and our scientific imaginings as activities within the world that we imagine" (Quine 1960, 5).

As straightforward as this idea may be, an embrace of it must entail a radical view of human objectivity. Indeed, the idea that we must understand our minds if we are to understand species, is every bit as radical as what Crick (1994) called *The Astonishing Hypothesis*, which is the idea that human consciousness has a scientific, biological, explanation. People, scientists included, tend to cherish their objectivity as an implacable part of themselves. It is part of what allows us to be scientists, and it may seem unsettling to turn our science inward and question ourselves in this way. But the evidence, our knowing failure to measure species, suggests that we must.

A Comment on Method

This chapter closes with a brief summary of some practical aspects of the way the central argument of this book is constructed. That argument is an explanation of the species problem. But up to this point all that has been done is to introduce that problem and to explain why we shall try a new kind of search for that explanation. Much of the remainder of the book is a description of that search. Yet no matter how fruitful this new search may be, it can only contribute one half of the explanation. The other half of the explanation of our species problem is a theory of biological diversity. Our understanding cannot be complete without a theory of what species really are, and of how they exist independently of our thoughts and language. It simply would not do to critique our language and thoughts without considering their subject. It would be like evaluating the design of a tool without considering its purpose.

Both of the major themes of the book, language about species and the reality of species, will be approached from a distance. The discussions of language and of biological diversity in the next few chapters begin in a way that is rudimentary and removed from the species problem. The reason for beginning with a reduced and detached perspective is the same for both language and biological diversity. Both are broad subjects, and in each case a direct embrace could easily miss key points that are best seen from a distance. For example, many discussions about language assume that real world referents have a reality that is as straightforward as they appear at first glance. Similarly, many species debates begin with a presumption that species exist, even though it is not exactly clear just what they are. Thus, on both fronts, language and species, we would like to avoid jumping directly into the middle of an existential tangle. Language can be examined in ways that does not entirely assume a particular relationship between words and aspects of the real world. Similarly, we can approach biological diversity by beginning with our more fundamental, and less problematic, knowledge of evolution. That is what we will do in the next chapter.

THREE

THE THEORY OF LIFE

The previous chapter described biologists' failure to measure species consistently, and how that persistent failure suggests a way to broaden our search for information that will help resolve the species problem. But still, even with an expanded search, the resolution of the species problem must lie within the domain of biology, the science of life. Even as we take a cue from investigators of human language and cognition, and allow our thoughts and speech to be included within the scope of our questioning, we must still recognize that imagination and language are processes of life and are just as biological as any other of our behaviors. Just as evolution has given rise to the diversity we see among organisms, so too has it given rise to the human capacities for imagination and language that we use to ponder the causes of biological diversity. This broad view of biology is not idle inclusivity. In this chapter, I begin a very basic description of life, and our understanding of it. In subsequent chapters, that description grows to include the causes of biological diversity and of our language about that diversity, and finally it will grow to where we have an understanding of the causes of the species problem. I recognize that some biologists would rather not search within human limitations for answers to the species problem, and this is one reason for stressing that we are not leaving the bounds of biology. But the more important reason is that the causes of biological diversity, and the causes of our limitations in assessing diversity, are mechanistically connected in the details of how evolution works.

As promised at the close of the previous chapter, we begin at points that are a bit removed from the primary topics of biological diversity and of our language about that diversity. In fact, the distance between initial arguments and final purpose is great enough that at first the connection between them will be obscure. This chapter lays out the rudiments of biological theory, and it demonstrates the role played by key words and their definitions within that theory. Neither the subject of species, nor the word SPECIES receive direct attention until later in the chapter. At that point we can compare our uncertainty regarding species and SPECIES with the uncertainty associated with other key biological concepts and words.

Our simple starting point is the theory of life. This phrase, "the theory of life," might seem incongruous at first, for biology is not always described as having a central theory. But biologists certainly do have a theory, one that is well grounded in evidence, that is simple, and that is widely accepted even if it is generally not acknowledged. Indeed, the theory successfully explains, albeit in very general ways, much of what we associate with life on earth. But in explaining life and its origins, the theory neither invokes nor explains species. Those theories that have been devised to explain the appearance of kinds of living things are uncertain afterthoughts with respect to the central theory of life.

The theory is not new. As can happen with old secure knowledge, biologists tend to assume the existence and truth of the theory without recognizing it as such. Its place in biology textbooks tends to be spread out among chapters on natural selection and chapters on the origins of life. It does not even have a proper name, and hereafter it is simply referred to as the theory of life. It goes something like this: (1) Somehow the physical and chemical circumstances of early earth gave rise to molecules that had some capacity for making copies of themselves. (2) As molecules with some replicative capacity multiplied, natural selection on replication took hold, and those molecules with a better replicating capacity became the ancestors of more molecules. (3) These early replicators may not have been nucleic acids, which were probably first employed by early replicators as aids to replication; but eventually replicators took the form of RNA, and then DNA. (4) A continual process of natural selection on DNA replicators gave rise to cells and organisms, the diversity of life (somehow), and our own capacity to observe and understand.

This version of the theory of life is short, but even much fuller descriptions of it are laid down with a very broad brush (Maynard Smith and Szathmáry 1995), well revealing that much remains to be discovered. The theory, unlike those in physics and chemistry, is not mathematical, but narrative, and vague at that. The details of steps (1)–(3) are woefully lacking, though they are the subjects of much investigation, and of entire research journals. Only step (4) could be regarded as unshakably grounded in evidence. Most biological research is not hindered by the uncertainty over the early steps because step (4) is not in doubt and this step began prior to the origin of the ancestors of all living things, all of which use DNA in very similar ways.

Some might protest that the theory is not what has been claimed, that steps (1)–(4) are a misleadingly incomplete theory of the origins of life. Certainly this is true regarding the richness of ideas and debates summarized in steps (1)–(3), but the only view I think would be found out of bounds with the descriptions provided is panspermia, the idea that molecular replicators first arose somewhere other than planet Earth and then somehow were delivered or fell to our planet. Of course, any theory of panspermia simply removes steps (1)–(3) to another local. Regarding step (4), some might object that other processes besides natural selection on DNA replicators must also be invoked. Certainly we could add that replication cannot always be perfect, and that there must be mutations for evolution to occur. I did not include it explicitly as it seemed implicit that when replicators arose they could not have been perfect, but it is indeed true that mutations must occur.

Another objection that has been raised in the past to descriptions of natural selection that are like step (4) of the theory, is that they overlook natural selection *at other levels* besides or in addition to that of molecular replicators like DNA. The main nonmolecule candidates have been individual organisms (the traditional favorite) and populations of organisms. This debate is best seen as a disagreement about where the DNA has an effect, about where the results of the expression of DNA are manifested. It is a debate over those consequences of DNA—out of all the possible consequences, including effects on the DNA itself, as well as effects on the organism that contains the DNA and effects on a population that contain the organism—that we think matter for natural selection. The DNA genomes of organisms exert an extraordinary array of consequences, and the building of the organism is but one (generally the most obvious one to our eyes). Consider several examples in which a mutation, a change in the DNA sequence, is favored by natural selection. First, consider that an altered DNA molecule that promotes the growth of a more durable organism may end up persisting and replicating because of that newly durable organism. Similarly, an altered DNA molecule that leads to a more reproductive organism will leave more descendant DNA molecules, in comparison to others, by virtue of the enhanced trait of organismal reproduction. So too may a change in DNA that promotes persistence or spreading of the population, that includes the organism that contains the DNA, lead to additional copies of the DNA molecule. In this case, the relevant effect of the DNA is manifested in properties of the population. Finally, we must not forget the simplest case, where an altered portion of DNA may be better at generating more copies of just itself, perhaps as a replicating transposable element or jumping gene. Such a transposable DNA may be part of the larger DNA genome of an organism, but even so it is possible for such elements to persist via replicating only themselves, even if the process of transposition is harmful for the organism that contains the transposable element. Transposable elements are extraordinarily common in the genomes of organisms. But regardless of whether the effects of the DNA are manifested locally just within that DNA, or as properties of the organism, or the population of organisms, the change in the DNA in each of these examples has improved the likelihood that the DNA will persist. Thus step (4) of the theory of life, which refers only to natural selection on DNA sequences, is a simplification but not an oversight.

A final objection is that the theory should refer to SPECIES and that it is simply not possible to speak of evolution without also speaking of species. This concern should not be confused with the idea that species necessarily evolve—rather, the flip side, the idea that evolution is necessarily, by its very nature, something that happens only to species. It is not difficult to envision that something like this might be the case. Consider that perhaps very early in the origins of life, evolution was associated with groups of related replicating molecules. Suppose further that these groups evolved into the modern species that we recognize (and disagree about) today. But such ideas are not generally discussed or included in the theory of life. SPECIES only rarely occurs in discussions about the origins of life or about the evolution that occurs among fairly simple replicators. For example, many experimental studies of evolution with simple organisms or with computer simulations do not resort to, or seem to require, a reference to simple species that are undergoing the evolution. Consider also that theoretical concepts of species, in which species are seen as the evolving units in nature, are not generally applied to the simpler molecular replicators that persist today, bacteria and viruses for example (Goodfellow et al. 1997; Van Regenmortel 1997), though there are exceptions (Gorman 1983; Zawadzki et al. 1995).

But the absence of SPECIES from discussions on experimental evolution and the absence of species theories for simple replicators could be seen as mere reflections of the larger species problem. Note that there do seem to be kinds of bacteria and viruses and it is a safe bet that had we watched the early times in the origin of life, that it would have seemed to us that there were many kinds of replicators. Perhaps a better theory of life would have such groups as central, functional entities. Yet, as much as we might imagine a theory of life that includes species —perhaps a theory in which the species is somehow the entity that is acted upon by evolution—there is not at present an accepted single theoretical species concept that would fit. I emphasize ACCEPTED and SINGLE, for there are indeed several species concepts that are consistent with the idea of early, evolving species of molecular replicators. There are also numerous species concepts that are theoretical and that invoke evolution, or contain the idea that evolution occurs to species. But by and large, these have not been devised with molecular replicators in mind. Thus, for example, the most widely discussed idea about species is called the Biological Species Concept (BSC). Under the BSC, a species is a group of interbreeding organisms that is isolated from other such groups. It is a highly explanatory idea for many organisms, but it is nearly pointless with regard to those bacteria and viruses that rarely engage in recombination. We will return later to the issues of species and groups of simple replicators.

Evidence for a Theory of Life

The claim for a central theory of life may seem odd, followed as it is here by discussion on its seeming crudeness and on the limited acknowledgment it receives. But at many points this book will rely upon that theory, and in this light it is fare to ask whether it enjoys much viability. Perhaps what was described above is a theory only in some narrow sense, and perhaps it is not used by biologists on a day-to-day basis. It is true that neither this theory, nor anything much like it, receives a great deal of direct acknowledgment. Ernst Mayr's rich history of the foundations of modern biology makes no explicit reference to anything like an overall theory of life (1982). This is so even though the book describes the discovery of the mechanism of natural selection and of DNA, and ascribes considerable weight and value to these discoveries; and even though the book is fairly riddled with discussion on the nature of biological theories, there is no mention of anything like the theory described here. However some other books are suffused with the theory (Dawkins 1982; Maynard Smith and Szathmáry 1995, 1999).

Ultimately, the evidence for a theory does not come from descriptions of it, but rather from the manner in which biologists proceed in their endeavors, in the way they wield their understanding in the course of their research. If not in the specific form outlined above, something like the theory of life persists in the minds and is implicit in the work of most biologists. For example, it can be shown that the theory contains, or at least implies, a penetrating, and highly reduced understanding of what it means to be alive. Modern biologists may not often speak of what it means to be alive, but they certainly act as if they have an understanding of the idea, and thus a theory of life. To help see this, it will be useful to explore the meaning of the word LIFE. It might seem paradoxical, but the theory of life does not obviously provide a definition of LIFE, and the word does not even appear in my terse summary until step (4). A casual observer might well wonder how we could have a theory about something when that theory does not even spell out just what that something is. But theories are about explanation and understanding, and these are not to be simply equated with definitions.

This point, that knowledge and understanding often do not lie in definitions, is an important component of the explanation contained in this book. The point applies well to SPECIES, as we shall see, but it also applies quite well to LIFE. To see this, let us pursue a definition of LIFE, and do so while avoiding the theory of life. Furthermore, let us be scientific and treat a definition of LIFE as a hypothesis, meaning that any particular definition is an idea that we are going to test, somehow, for its suitability. Furthermore, we are simply seeking a definition in a conventional sense, and that is a description of the property or set of properties that all living things must uniquely have.

Perhaps a good starting place is the cell theory of Schleiden and Schwann, which holds that a living thing arises from a cell and is made up of one or more cells. This 19th-century idea cannot compete with the theory of life, at least on a mechanistic level, but it is a wonderfully simplifying notion. The cell theory says nothing about the origins of life, nor even of exactly what a cell is, but beyond that it is wonderful for its nearly complete accuracy. Yet there are things that fit virtually all other definitions of life, and do not fit the cell theory. In the first place are viruses, which are little parasitic packages of nucleic acid wrapped in protein capsids. Viruses can certainly grow and reproduce and evolve, though they are not considered to be cells. Perhaps one could argue that a virus is a tidy bundle and thus a type of cell, broadly defined. Another exception comes from animal embryology. For many animal embryos there is a stage when the complete embryo, or a portion of it, forms a syncytium in which there are no cell membranes between nuclei. The syncytium does not have cells, yet would presumably be thought alive by almost any other notion. Similarly the appropriately named acellular slime molds usually exist as a plasmodium, a kind of sack of protoplasm and nuclei. If we are to be strict about a definition, then we must reject the cell theory definition of LIFE, even though it fits most things that we commonly think of as living.

Another possible definition is that LIFE means having the capacity to undergo a process of metabolism, a regular flux of energy and matter, despite a relative constancy of form. But this criterion also applies to some distinctly abiotic things, such as a flame (Maynard Smith 1986). It also does not apply well to viruses. Virus particles have no flux of mass, even though they have evolved and can grow and reproduce as parasites within host cells.

We turn then to another useful idea, that LIFE should be used to refer to an instance of something having the capacity to grow and reproduce. But, of course, there are crystals, which also grow and can even reproduce in a fashion, though not quite in the way we think of cells or DNA molecules reproducing. Another problem with this definition is that it does not apply very well to older organisms beyond their time of reproduction. Perhaps this kind of exception is not too much of a problem as it is true that such organisms still consist of cells, many of which are still growing and dividing, and so we might say at least that they are made up partly of living cells. But what to say of mitochondria and chloroplasts that grow and reproduce within cells? Perhaps we can admit that these cellular organelles are alive, even though it may be a stretch according to more everyday thoughts on what it means to be alive. But here is another, uniquely modern, exception to this particular definition. In biological laboratories, it is an easy and routine procedure to create a test tube environment in which strands of DNA grow and reproduce. The most common manifestation of this is the Polymerase Chain Reaction (PCR). This example does present a dilemma for the definition, for we must either conclude that the DNA is alive or that the definition fails.

One more definition that we should consider is that living things are those that are capable of evolution, a process that requires primarily the capacity to replicate and leave modified descendants (Maynard Smith 1986, 6). This idea is closely related to the point, perhaps first made by Muller, that a replicating gene must be the basis of life. Impressively, he made that point in 1929, long before we had knowledge of the material nature of genes (Muller 1929). Yet this definition would also apply to mitochondria and chloroplasts, as well as some naked DNA molecules in the laboratory that can be made to evolve using circumstances similar to the PCR reaction. This definition would also apply to memes, which are simply ideas that can be passed on from person to person (Dawkins 1976), and indeed to anything with some inherent tendency to cause its own replication. Computer viruses certainly are alive under this view, as are chain letters that evoke assistance in replication from trusting humans (Goodenough and Dawkins 1994).

This exercise of pursuing an unambiguous definition of LIFE has failed. Consideration of various definitions has shown that regardless of our tack, there arise circumstances that prevent us from drawing a clear line between things that are alive and things that are not. We could use Maynard Smith's definition, since those unambiguous living things— such as cells and organisms, which fit all of the definitions—clearly arose by evolution and persist by evolution. But then how to decide about other things that evolve, like computer viruses and memes. By the criteria of evolution, a decision to exclude them will be arbitrary. A part of this difficulty is how to decide when something that we might agree is sometimes alive, is not alive. For instance, if we consider a strand of DNA to be alive because it can replicate and evolve under some laboratory settings, do we also consider it to be alive when it is stored in the freezer or dropped on the floor? If we admit an idea to be alive, is it alive when it is contained only in a book or a buried scroll? If in each of these cases we were to say that some static thing may have the potential to be alive, but is not alive, then how do we decide at which point it becomes alive? Summing up our dilemma, we see that every definition has the difficulty that it is arbitrary in initial delineation; at the outset arbitrary lines must be drawn. So, too, in application, can we expect each of the definitions to suffer vagueness. No matter our definition of LIFE, there will be borderline cases when our definition is not a useful guide for deciding if the word is applicable.

The exercise has also revealed a flawed starting point, at least in so far as I, or a reader, might have assumed that an unambiguous definition could be found. Such an assumption would certainly be reasonable given the way that many definitions work. Certainly it is reasonable to suppose, not just for LIFE but for many words, that some distinct set of circumstances have given people cause to come up with the word, and that these same circumstances could be used to generate a complete description of what is common to those cases where we use the word. Yet even though LIFE is a common word and even though generating definitions for words is a common enough activity, we are unable to generate a precise definition. Certainly the word can be lovely in usage; it is usually clear in meaning in so far as both the speaker and listener feel they understand, and it is often wonderfully evocative. But we seem to be laboring under a historical burden. A word that arose in simpler times has become vague in light of modern understanding of the diversity of living things, and the properties of some things that may or may not be living.

Now let us place what we have learned about the meaning of LIFE into the context of the theory of life. That theory concerns the origins of life, and built into it is the explanation of why the word LIFE is vague. The theory simply says that living things arose from nonliving things, and it neither presupposes nor imposes a distinction between the two. It is simply a description of how some present-day things probably had their origins in some molecules that long ago began a process of replication. It is a shortcoming that we do not know the chemical details of how the first replicators got started, but we have every reason to think it did happen.

In practice, biologists are not troubled by the ambiguity of LIFE and they generally understand that even though we are alive and even though we work on living things, that living things do not all have a unique property that sets them distinctly apart from all nonliving things. Perhaps the clearest indication of this kind of understanding is that no modern biologist has an interest in creating life per se. There are many scientists who are in the business of trying to create a certain kind of life, such as a very simple cell or a computer program that mimics biological phenomena. But life per se is not their focus. Biologists' willing vagueness about LIFE is a clear sign that they think they have some kind of serviceable set of ideas that account for the existence of life. Suppose that we did not in fact have this kind of understanding. Then we might expect that the meaning of LIFE would still be a debate headliner. In fact, research papers on the origins of life are not at all devoted to the meaning of the word, though they do chew furiously on the details of the kinds and origins of replicators. Today, if you hear a biologist making claims about creating life, you can be sure of one thing: that they do so with an explicitly, and arbitrarily, narrow meaning of LIFE.

The evidence then, that biologists have a working theory of life, lies in the way they do not suffer for their uncertainty over words and their lack of crisp definitions. The word LIFE may be the most extreme example, but there are others that are actually more pressing on a daily basis. In each of these cases, vagueness and uncertainty may pervade the language, but the scientists are tranquil, confident in their understanding of the causes of those things we give words to. Consider three of the most widely used words in modern biology: CELL, ORGANISM, and GENE. All of these terms suffer vagueness, but scientists and scientific progress do not suffer because of that vagueness.

CELL is actually about as unambiguous a word as one is likely to find in biology. Unlike many of the things that words are used to refer to, real cells are well-bounded entities. Indeed, "well-bounded entity" could arguably be considered as a definition of CELL, though one broader than the biological meaning of them word. But there still do arise occasional circumstances where it is not clear whether CELL should be applied, as mentioned above, particularly if you would follow Schwann and Schleiden and insist that all living things are made of cells.

What is the cost of this ambiguity? Imagine a cell biologist introducing a new student to the ubiquity of cells across life. The lesson might include discussion of the cell theory as well as discussion of the importance of understanding the way cells work as a necessary step for understanding how organisms work. The student might rightly think that the cell is a truly fundamental feature of life. It would be understandable if she went even farther and thought that life could not do without cells and that any fully explanatory theory of life would have CELL in a central role. Suppose that on the next day, the student learns of the exceptions—the acellular slime molds, for example—and brings this to the attention of the professor, thinking it to be evidence against some grand theory. But any modern professor would probably respond with something like "Yes, isn't that fascinating," with no concern for the status of a grand theory. What would such an exchange imply? Just that the teacher has a better theory of life than does the student, one that does not rely on cells. In brief, the understanding held by the teacher in this example, and of that held by biologists in general, never suffers from the occasional uncertainty over whether or not CELL fits a particular situation.

Now consider ORGANISM, a word that is arguably even more fundamental than CELL, tied as it is to some of our most basic ideas on individuality (Wilson 1999). But there are a few well-known circumstances in which we do not have a good way to count or identify organisms, and where ORGANISM is vague. Symbiotic amalgamations, like lichens, grafted plants, and people with transplants present a bit of a muddle. Another is presented by sponges, which are made up of cells that can be separated and then brought back together to approximately reform the organism. When the cells are separate, how many organisms are there? Another kind of organismal ambiguity occurs as a colony of eusocial insects, such as a honeybee hive. The colony certainly includes many organisms, but the colony itself is a kind of organism (Moritz and Southwick 1992). Finally consider the boundary of an organism, the edge between it and the outside world. Many organisms have attached portions of secreted material or dead tissue. Do we count these extremities as part of the organism or not? If we do think of, say, a clam shell as part of the clam, then would we not also count the nest of a sea swallow, which is made largely from secreted saliva, as part of the sea swallow? In short, organisms construct and shed their boundaries in such a variety of ways that it is not always clear what part of the boundary should be considered part of the organism.

Note that one thing we have definitely not done by uncovering vagueness in these linchpin words of biology is uncover holes in our understanding of living things. The ambiguity does not arise because of a lack of understanding; rather, the reverse, the ambiguity is revealed by a thorough understanding. The ultimate source of our willing, occasional uncertainty about CELL and ORGANISM is our knowledge of the historical process that gave rise to life. Living things evolved from nonliving things, and there is little reason to think that any particular aspect of living things is truly necessary, that we could be completely assured that it must be shared by all living things or that it would evolve again in some replicate of the origin of life on earth. So far as we know, the theory of life does not predict that cells or organisms are either necessary or necessarily unambiguous when we find them.

Considerably more vague than either CELL or ORGANISM is GENE, though it did not start out that way. In the midst of the confusion over heredity that followed the rediscovery of Mendel's laws, early in the 20th century, Wilhelm Johannsen was prescient in his careful construction of operational definitions (Dunn 1965; Provine 1971). First he saw that researchers were suffering from confusion between the information for the organism, that is contained in the organism and is passed from parent to offspring, and the actual organism. Johannsen's two famous words that enable the distinction are GENOTYPE (the genetic constitution of an organism) and PHENOTYPE (the sum total of an individual organism's expressed characters). He also foresaw the need for a way to refer, however hypothetically, to a unit of the genotype; hence GENE. Since 1909, when it was coined, the word has been a central pillar in virtually all discussions of inheritance, with the core idea being that a gene is the unit of the genotype that encodes a distinct function found in the phenotype. At times this idea has seemed to nearly become a reality. For example, from the success of the chromosome model of inheritance, and the success of gene mapping studies, we know that the idea that genes are distinct entities strung along a chromosome is approximately correct. But today it is also clear that the function of DNA sequences takes many forms, not all of which have sharp boundaries or are separate from one another. An incomplete listing of functional categories includes DNA sequences that code for RNA molecules (some of which then code for protein and some of which do not), intron sequences that code for correct intron splicing, promoters, enhancers, and telomeric sequences. Furthermore, the boundaries for some of these functional sequences, in terms of their location in a DNA sequence, may not be distinct, and there are many cases where DNA sequences associated with one function are part of a larger sequence that is associated with a different function (e.g., enhancers lying within protein coding regions). Thus a gene can be hard to pin down, and this is for the simple reason that the information in the genotype that corresponds to function does not always occur as distinct segments of DNA. Today GENE is probably our single most common biological term. It is a bit ironic that as our knowledge of the genetic material has grown, GENE has become more difficult to define. If one relies on a narrow clear definition, such as "a region of DNA that codes for a protein," then there are other critical, functional parts of the genome that would be excluded. If one prefers a broad definition, such as "a functional unit of the genome," then one must also accept that some of those functional units will overlap others and some of them will not have sharp boundaries.

When it was coined, GENE had an unambiguous precise meaning, although that meaning was entirely theoretical. The definition and application of GENE was precise simply, indeed precisely, because it was theoretical. The word was an invention, and at the time nobody knew how that invention might correspond to something physical in nature. It is an example of an ironic but common kind of switch that comes with understanding. As we learn more about something, the more our word for that something becomes ambiguous. We do not suffer the ambiguity, generally, for we have a broader understanding. We can explain phenomena, but if pushed for a precise definition that is also widely applicable, we may be stuck.

GENE, CELL, ORGANISM, and LIFE are fundamental words, but only in the sense that they are commonly used, and are used in reference to common aspects of nature. They are not fundamental in the sense of

being completely unambiguous. Nor are they fundamental components in the basic theory of life that was outlined above (though they
are of course mentioned in theories that try to explain them). Yet these
words are very widely used, and are very explanatory; the vast majority of the time they are used, their meaning is conveyed without error
from speaker to listener. How could we have words as integral to biology as these and yet not suffer for lack of precise, unambiguous definitions? The reason, again, is that as important as these words are, they do
not form the foundation of biological theory. The words are important, but not *that* important. I would conjecture that there is not a single context, wherein one of these words is ambiguous, that biologists
cannot provide an explanation and an understanding of the processes
therein. Underlying all of our understanding of complex aspects of living things is a theory of the origins of life.

Theory, Words, and Species

What does the occasional ambiguity of linchpin words in biology have
to do with the species problem? The answer has two related parts. First,
they show us some things about how understanding in biology works
when there is a good theory. We find that even when biology is at its
most basic and most certain, that certainty is not contained in unambiguous definitions. This does not mean that clear, simple definitions
are not useful tools. They are exactly that, as every teacher and student
knows. But the definitions are not the primary containers of knowledge. Rather, they are like small disposable cups: useful for quickly and
easily scooping portions from the larger pot. Definitions fail at borderline cases, yet in those same cases a broader knowledge may stride pain-
lessly. It is precisely because we do have a full understanding of life, and
of the possible definitions of LIFE, and of the reasons why we have borderline cases, that we don't suffer any sort of LIFE problem as we do a
SPECIES problem.

Second, we can compare this rosy picture, that fits most of biology,
to the malfunctioning discourse that surrounds the species problem. A
curious byproduct of the species debate is the large number of definitions that have been proposed for SPECIES. Many biologists have wor-
ried, debated, and argued about the best definition, and some have
spent a considerable part of their career carefully fashioning, and then
honing in response to criticisms, terse "perfected" definitions. I have
participated in this as well, and have circulated a manuscript that de-

scribed yet another species concept. Some 24 different species concepts were cataloged by Mayden (1997), and now that list is incomplete (de Queiroz 1999). We have, in this corner of biology, a strange industry focused on producing a lengthy dictionary for just one single word. Given the persistence of the species problem, and given what we know about the modest role played by definitions for other key biological words (such as LIFE and GENE), it is fair to wonder whether SPECIES pundits have been paddling around in a semantical abandoned meander, caught in some sort of misunderstanding about what a correct definition of SPECIES could do for us. I think that is exactly what has happened. Biologists do have a good theory of life, and we have some very good ideas about the causes of biological diversity, but we are unsure of just how to reconcile our perception that there are kinds of organisms, with that theory, or even if we should try to do so. In our uncertainty, and faced with demands from society for counts and descriptions and names for species, we have tried to fashion a fix by finding exactly the perfect set of words that will cause things to fall into place. As we shall see, this cannot work; our species problem runs deep, and it is not to be purged by any short sequence of words.

What does the theory of life have to do with the species problem? Quite simply, we can use it to develop a theory of species, or rather to bring species into that theory. The theory described near the beginning of this chapter is a stripped-down starting point from which we can build an understanding of why we have kinds of organisms.

Part I Conclusions

Here closes the introductory portion of the book. The purpose of these first three chapters has been to describe the species problem, and to set the stage for developing an explanation of its cause.

Biologists well know that their field has, for decades, suffered obdurate debates over the identification of species and over the optimal definition of SPECIES. The beginning of a way out of this morass lies not in continuing those debates, but rather in consideration of a largely untapped body of information, which is our own behavior in the course of debating species and SPECIES. In chapter 2 we saw how biologists persist in making subjective and irreproducible measurements of species and that we do so in the face of great experience of that subjectivity and irreproducibility. In chapter 3, we saw how ideas of species, including definitions of SPECIES, fall outside of mainstream ideas of the theory of life's origins. In particular we saw how our uncertainty over species concepts is not at all like the uncertainty that comes with other key biological concepts. Our behavior over species and SPECIES is, in some ways, irrational, and we could well promote our understanding of biological diversity if we could understand the cause of that irrationality.

The argument that is developed in part II has one component that is clearly biological, and that builds upon the theory of life described in chapter 3. The second and more novel component of the argument follows from an investigation of human thought processes and, particularly, language, as they occur in regard to species. Language is the subject of chapter 4, and it is there that the pursuit of the cause of the species problem begins in earnest.

Species in Nature and within the Mind

FOUR

CATEGORIES

Is it possible that human minds and human language are dysfunctional in some way that causes them to be poor tools for understanding and describing biological diversity? This is the possibility that was inspired by the repeated knowing failure of biologists to measure species, and that was discussed at the end of chapter 2. In chapter 3, we stepped back and explored the theoretical basis of biological understanding, and the role played by some relatively straightforward words (like GENE) within that understanding. At that level, which stopped short of species, we found no noteworthy conflicts between language and our best understanding of reality, despite the fact that common words are sometimes ambiguous. Now we turn to focus more explicitly on language and on the way that language corresponds to reality. People use language to further a seemingly endless variety of goals, but is it an inherently reliable tool for building a model of the larger world? And if it is not perfect, why not? And where lie the imperfections? By asking, I assume the question is sensible, and thus also assume considerable utility on the part of language. Necessarily I presuppose that language is just great for many things, even for communicating about language. But language is complex, and the real world of which language is a part is also complex, and the fit between language and reality could be better in some respects than others.

Most scientists do not often broadly question their words or the relationship between language and the real world. Their innocent speech and text is at least reasonable, for a person cannot simultaneously critique all aspects of their own inquiring process. It is also reasonable given the more expert view held by the smaller community of scientists and philosophers who do explicitly question the relationship between language and the world. For them there is a sort of bedrock principle of tight correspondence. "There is a long philosophical tradition according to which the basic structure of language and the basic structure of reality are the same. The basic linguistic structure is supposed to be that of subjects and predicates while the basic ontological structure is supposed to be particulars and universals" (Martinich 1996, 183).

This idea is an assertion that language does fit the world; but more than that, it is a specific description of structural commonalties: subjects of sentences are to particular entities, as the predicates of sentences are to universal classes or properties. In "My car is red" the subject CAR refers to a particular entity, and the predicate RED refers to a universal property. In "The sun is a star," SUN refers to a particular star and STAR refers to a universal category. This view of the way that language and reality match up, which I will hereafter call the *common structure model*, has stood the test of time in a sense, as it is not generally questioned directly in the works of philosophers and linguists. But perhaps because the common structure viewpoint is a bedrock principle, there does not exist a strong tradition of questioning the quality of the fit between language and the world. Please understand my use of QUALITY in this last sentence, so as not to confuse it with the long philosophical tradition of inquiry on what language contains—on issues of meaning and truth and on the reality of what words refer to. Philosophers have indeed inquired extensively of those things. But in saying that the *quality* of the correspondence between language and reality has not been studied, I mean that there has been little direct questioning of the circumstances when language is better, or worse, or of the reasons why it is better sometimes than others. These last questions are our concern. For the reasons laid out in chapter 2, there seems the possibility that one cause of the species problem may lie in the way that language about species works; that something about the language process is ill suited to the reality of species.

Of course my phrases like "the way that language works" or "circumstances when language is better," as well as my frequent use of LANGUAGE are bits of language, and not very precise bits at that, even in their full context. Words about words are awkward at best, and circuitously vacuous at worst. Language is all tangled up with thought and understanding and communication and a host of other related human processes. And as bad as these large tangles are zillions of smaller semantic ones—note that not one of the nouns in that previous sentence is easy to define. Could we possibly puzzle out the quality of language, or describe the results of such puzzling, using language? Some hope that the language tangle per se will not prevent an inquiry comes from the history of study of language, and indeed of science in general. Just as did all those who ever tried and succeeded to make sense of anything, language included, we must assume that language can convey thoughts and have meaning and some correspondence with the real world. We must be both questioning and accepting; willing to take many statements at face value in order to critique others. Language need not be perfect for such a process to work. This was Quine's point at the beginning of *Word & Object*.

Where shall we begin? Actually we do not need to look far for one kind of conflict between language and reality, as we have already touched on some ways that language does not fit the world, at least as stated by the common structure model. That model supposes that the discrete cleanliness of a simple sentence, like "The governor is Catholic" matches an orderly reality in which, for example, real governors either are or are not Catholic. But, of course, sentences must be literally discrete and well bounded, for we can make little use of a fraction of a word or of a letter. In contrast, there is no necessity for the real world to be so well packaged. Recall the example of clouds in chapter 2, some of which are much more distinct than others. Many species as well are fuzzy in that they do not have distinct boundaries. A great many features of our world are like this, as we well see the moment we stop to notice. Mountains seem distinct at their tops, but not at their bottoms. Even the earth is indistinct, with an atmosphere and Van Allen belt that diffuse gradually into space. But language, whether spoken, signed or written, is essentially digital. It is made up of a finite set of discrete components, symbols or sounds, and it carries information primarily in the sequence of those components. Since language is discrete, and much of the real world is continuous, there are necessarily a great many circumstances where language can only be an approximation (Watzlawick et al. 1967). We could imagine that in the limit, with infinitely long sentences, we could perfectly describe a fuzzy part of reality, just as digitally sampled music can be completely faithful to the continuous original only when that sampling is carried out to an infinite degree. But we are not capable of such perfectly long descriptions. More important, we do not condone long descriptions—in practice, terse, discrete language is fitted to a fuzzy world in some approximate fashion.

Could the mismatch between discrete language and fuzzy reality be the cause of our species problem. The disparity can confound us at times and no doubt does contribute to the problem. But when we pause to notice how many aspects of the world are not distinct, we readily appreciate that the simple contrast between crisp speech and blurry reality cannot be the main source of our difficulty. If it were, then we would also have lots of trouble using language for all those other aspects of the world that are fuzzy. But we don't have a cloud problem, or an earth problem, or indeed anything else quite like the species problem. Sometimes words for things that are not always distinct do generate debate and discussion about when the word should be used. For example, DISEASE is among the more vague biological words in biology, and it is also a highly necessary word. The vagueness and necessity surrounding DISEASE inevitably lead to questions and debate on the meaning of the word (Caplan et al. 1981; Thagard 1996). But as interesting and valuable as such DISEASE debates may be, they are not contentious and long-lived the way species debates are, nor are they both broadly conceptual and narrowly obsessed with short definitions, the way SPECIES debates are.

What other aspect of language deserves scrutiny? Posed so generally, the question suggests a hopeless needle in the haystack dilemma. Language is far too multifaceted and complicated to be generally deconstructed and critiqued in any short space. Fortunately, we can take hope from the circumstances of our particular pursuit, which is much more limited than a general critique. Rather, we have some concerns about the way language is used regarding species. Even better, our limited concerns connect fairly directly with a distinct, and strong, tradition of inquiry on language. That tradition concerns categories, also called classes or kinds. Philosophers have long questioned the nature of categories, both as features of the world and as features of language, and this philosophical tradition has seeded a chain of questioning and research about categories that runs through the modern fields of linguistics, anthropology, psychology and what is today sometimes called cognitive studies.

Categories are so central to our speech and thought that to even think about them directly requires a fair bit of mental effort. Sometimes they seem simple, such as in "The baby is a girl," where GIRL is the name of a category. But note that in the sentence "GIRL is the name of a category," the word CATEGORY is a category—it is the category of all categories. As Lakoff notes (1987, 5–6), "Categorization is not a mater to be taken lightly. There is nothing more basic than categorization to our thought, perceptions, action, and speech. Every time we see something as a kind of thing, for example, a tree, we are categorizing.

Whenever we reason about kinds of things—chairs, nations, illnesses, emotions, any kind of thing at all—we are employing categories."

But what are categories? The question has occupied thinkers for thousands of years, and I will not address it entirely. But there is one thing we can say about categories, and that we can use as a starting point: a category is a word (or an idea in the mind) that we use to help organize other words (and thoughts) about recurrent features of the world. This claim has three key parts that deserve reinforcing. First, and foremost, a category exists in the mind of a person, and can be expressed in speech (and other modes of language expression). Second, a category is a major component of an organizing, or classification, process. Third, a category is for those times when there is recurrence, when similar circumstances, or potential similar circumstances, are being considered.

The existence of a rich tradition of thought on categories means that we can, in short order, draw upon a great deal of information about how references to species occur in language. The reason is that species, whatever else they may be, are kinds (categories) of organisms. Please note that I am not now presenting a definition of SPECIES. By saying that species are kinds of organisms, I am simply pointing out that people regularly construct sentences with a syntactical structure in which a species name appears as a category. Statements in which an organism is identified as to kind (e.g., "I am a person" or "She was bitten by a dog" or "He ate a lobster") are very common. Furthermore, in many of those sentences, the category, the *kind* of organism, corresponds in some way to what we might, by some other criteria, call a species.

Realism and Categories

The traditional Aristotelian view of categories is that they are each characterized by a property or set of properties that are necessary and sufficient for membership in that category. The list of properties forms the essence of a category. This view is a tidy one and corresponds wonderfully to the basic mathematical and logical idea of a set. The framework works especially well for categories that we delimit in our descriptions so that, for example, the RED THING category includes just those things that share the property of redness. The framework also partially works for discovered categories, also called natural kinds, which are categories that seem to appear to us as recurrent aspects of nature. Natural kinds are the categories that are of particular concern to scientists. Every scientist is deeply concerned with some natural kinds, their subjects of study (e.g., tornados, fish, gravity, numbers, neutrinos, heart attacks), and thus with the shared properties of the things that seem to belong to the natural kind that is their focus. The very many different natural kinds (tornados, fish, etc) seem to share some interesting properties that set them apart from other classes, such as those we invent or delimit. In the first place, the different instances of any one natural kind seem to share a great many features, and the more one investigates the more new common features tend to emerge. It is as if the tabulation of defining properties could go on and on the closer one looks. Second, further study of members of a natural kind usually turns up causes for why so many common features seem to go together. Thus, for example, tornados are a class of storm with all kinds of properties that set them apart from other kinds of storms, and from research on many instances of the TORNADO category, the processes underlying the commonalties have been gradually revealed. These many commonalties shared by particular tornados, or of instances of any natural kind, are the brickwork for human reasoning by induction—the often hidden process by which we make assumptions about an entire category of things, by reason of what we know about similar instances of the same kind of thing. Induction is sometimes defined as reasoning from the particular to the general, and it is not usually considered to be a scientifically rigorous mode of reasoning. The criticism of induction as non-scientific is that we cannot fairly say that we know something to be true for all members of a natural kind, just because it happened to be true of some members. Be that as it may, induction is the business of using knowledge about what has happened in the past to make predictions about the future. Furthermore, when it works and predictions are confirmed, it can be taken as evidence regarding the properties of a natural kind, and as some sort of affirmation of a natural kind.

The continuous revelation of properties of natural kinds, as they are investigated, and the success and utility of induction regarding natural kinds, have fostered debate over whether natural kinds are real and exist. At first glance it might seem like a silly debate. I said that tornados and fish are natural kinds, and who would question the fact that tornados and fish exist? Why would I bother mentioning them if they did not? But blurred within these last few sentences is a critical distinction, a world of difference between two ways that we may use a category. I can use FISH to refer to some particular fish (e.g., "the fish that are living in the ocean") or a particular fish ("the fish I ate for dinner"), and such particular instances may indeed exist. But I can also use the word FISH more literally as the name of a category. It is precisely in this sense, the category-as-thing sense, where existence is a puzzle. Does the fish category exist?

Certainly the fish category, and every other category we might name, does exist in one particular way. I can name and describe a category, and I can use it in speech and text, and so at the very least a category can exist in my mind, and in the minds of others who use that category. Note that as mental entities, categories are not always the well bounded things that we often assume them to be. For example, I might insist that the fish category should be applied to just those aquatic vertebrates that have fins and gills; and perhaps you would agree. If so, then what are we to make of all the other things that I identify with FISH? The FISH category in everyday use includes real living fish, past and future fish, some unseen things that pull on fishing lines from beneath the water, pictures of fish in books, images of fish on TV, and fish in dreams—some no doubt rather strange. As much as the defining criteria of the category might seem to delimit the things for which we would use a category, the actual ideas and concepts that are our categories are often deployed with a great flexibility. In general, a natural kind may seem orderly and well bounded by shared properties, but the category itself resides in a person's mind and may be used in nearly boundless ways, ways that are delimited more by human creativity than the actual properties by which a category is conventionally defined.

But can a natural kind be more than an idea? Does the FISH category, or any natural kind, also exist apart from ourselves? There has been a lot of philosophical debate on the existence of natural kinds. Those who are especially impressed by the continuous unfolding of multitudes of discovered properties, and of the utility of induction, often hold that the essence of a natural kind does have some ontological status (i.e., the kind exists in some way). Traditionally, this view is called realism. Opposed to this kind of realism stand a variety of viewpoints (Hacking 1983; Putnam 1981), but for the most part we can focus on just the most basic, what is sometimes called the nominalist critique. The main idea behind the nominalist critique is a kind of plain talk, no-nonsense view of existence. It is the idea that individual things exist but nothing else does. The nominalist critique draws a hard line between a category (which does not exist), and an instance of that category. An individual (i.e., a particular entity) exists, and a category can include individuals, or other categories, but is itself an abstract construction that does not exist. The critical difference between entities and categories is that the former are concrete (tangible), whereas the latter are abstract. Thus one can throw a stone because it is a physical entity but one cannot do anything to the STONE category. Another distinguishing feature of an entity is that it is localized in space and/or time, within some restricted part of the universe, and does not occur just anywhere or everywhere. In contrast, whatever existence a category might have outside of the mind, it cannot be pinned down by location in time or space.

As simple as the nominalist critique may seem, there still arise a great many uncertain situations, most of which have to do with what is, or is not, an entity. I said that an entity is concrete or tangible, and in their most obvious sense these criteria serve well for particular fish or for particular stones. But what about the many other things that we refer to, like particular relationships, that seem to exist and are not categories, but also are not simple entities? Does a shadow exist? Does a marriage exist? One thing that helps in these cases is to expand the conventional meanings of CONCRETE and TANGIBLE to include, not just things that you could hit with a hammer, but also all those things that exert effects or that can be acted upon. You may not be able to physically hold a shadow or a marriage, but a shadow can affect the growth of a plant and it can be altered, and a marriage can be affirmed or broken. Note too, that a particular shadow, or a particular marriage is somewhat localized in space and time.

I have made repeated examples with FISH, which is a name for a kind of organism (though a larger kind than a species—there are many species of fish). I could have used most any commonly recognized kind of organism, for kinds of organisms are quintessential natural kinds. In this light it is not surprising that part of the historical debate over species has been a kind of reenactment of the realism/nominalism debate. Mayr (1982, 264) describes the nominalist, antirealist stance of some 18th-century biologists and some 20th-century writers have taken this tack as well (Burma 1954; Gregg 1950) and argued that species are categories and thus do not literally exist. Another view fully accepts the nominalist stance regarding categories and natural kinds, but also says that species are real entities and are not natural kinds at all (Ghiselin 1966, 1974; Hull 1978). This is an important point, and we will return to it in various ways, particularly in chapter 8.

To our questions regarding how words fit the world, the realists and their critics have different takes on the common structure model. If I say "the earth is a planet," then either my PLANET category corresponds to some real world natural kind that exists outside of one's mind (the realist view), or my word is at best a convenient device for referring to recurrent circumstances in nature, a kind of shorthand with no special mapping onto a real entity. To a realist, language can be a literal model of reality with both the subject and the predicate of a sentence serving as symbols for parts of nature. The nominalist critique is not traditionally explanatory on what categories are, except to say they do not correspond to real aspects of the world.

For the many scientists who are concerned with concrete entities, consisting of mass and/or energy and that are at least somewhat localized in space and time, the debates among philosophers over what is real may seem irrelevant. If faced with arguments that existence is not limited to tangible things, but also includes some abstract things, then they can simply re-explain their interests in terms of those features of the world that are concrete and avoid an ontological lingo. If asked to grapple with this debate, probably most scientists would request a transfer and choose to sidestep the ontological uncertainty of possible real, yet abstract, things. This pragmatic stance is similar to the one that I will take, as I hold out a primacy for the reality of tangible entities. In most of the discussions of biological processes to follow, I try to consider just the concrete and tangible aspects of nature, including just those parts of nature that can change without regard to our perceptions or words about them. It is important to note that in taking such a practical stance, that one also tacitly accepts the nominalist critique of categories. As a practical matter, natural kinds are at least effectively not real. Even if a good case can be made that in some ways they are real, this would be a different kind of reality from all those particular happenings of mass and energy that are localized in space and time, and that can come and go and change.

The Overturning of Categories

But the philosophical question of the reality of natural kinds was not brought up just to be sidestepped. The long history of philosophical inquiry over natural kinds has seeded a strong, modern tradition of the study of categorical references in language. As a way to introduce this, consider the implications of the critique of realism. If that critique is correct, then categories do not literally exist in the world outside of ourselves. Rather their place is within us, in individual minds and in communities of minds that exchange words and ideas. Thus one of the most direct implications of the nominalist critique is that one must do psychology if we are to understand categories.

Let us now follow a different approach to categories, one with a psychological focus, and let us start by bringing this focus to bear on the classical Aristotelian view, in which categories consist of suites of shared properties—rules, if you will. Is it possible that this classical view somehow corresponds to the way our minds actually construct and use categories? Though it was devised by philosophers and not psychologists, the classical view can be used to construct a rough psychological rule-based model of how the brain might represent categories. Suppose, for example, that the brain maintains a list of rules, in association with each categorical word or idea. Perhaps the categories in language are maintained in the brain as a kind of tabular database, with each category represented by a list of properties. Certainly a computer can be programmed to categorize things using a set of rules, and so perhaps our brains do that as well.

In recent decades, a variety of investigators from various fields have questioned categories, and when these various insights are brought together it does not seem that the classical Aristotelian vision describes them very well. For example, Wittgenstein (1968) showed that some words are used for categories of things that seem to have no defining properties. His famous example is the category of games, which includes among other things: chess, Russian roulette, solitaire, and ring-around-the-rosy. They are all games under one general meaning of the word (i.e., they are not homonyms), and they share what Wittgenstein called a family resemblance, but it is hard to say why they are all games. Other examples are not hard to come by; consider garbage which as a category commonly includes, among other things: egg shells; evil people; poor writing; and pornography. Perhaps the point is best demonstrated by reading a dictionary. Take notice of how, in order to get a good sense for when a word might be used, it is often not sufficient to read just a single definition. For many words, even simple categorical nouns, one must read several closely related, but partly distinct definitions, to get a feel for usage.

Evidence against a rule-based mental method also comes from work by psychologists on recognition and memory. In a now famous series of experiments, Rosch and colleagues showed that people do not have a mental representation of categories that fits with the classical view. Rather, people use categories as if there are good and bad—or central and peripheral—examples. Consider the BIRD category, and notice how some birds seem much more typical of the category. What leaps to mind at the symbol BIRD? If you are like most people, your mind first generates an image of a bird you've often seen, perhaps something sparrow-like, rather than that of a penguin or an ostrich (even though you know that all are birds). Or consider the CHAIR category. When you interpret the word CHAIR as a reference to a kind of thing, does a list of criteria run through your mind, or does a prototype pop up in your mind's eye, a kind of generic but quintessential chair? If it is the latter, then you would probably agree that some chairs seem more chair-like than others. A stool, or a large bean bag, or a child's car seat are all things with a purpose of being sat on by people, but they probably all seem to be peripheral chairs compared to a rump-sized four legged platform at knee height, with a vertical back rest. The existence, within the mind of some sense of typicality for many categories is also manifest in the results of a wide range of psychological tests. For example, the speed at which people can classify items is greater for items that more closely resemble those that have been rated as most typical for a category (Rosch et al. 1976).

From a synthesis of diverse aspects of the ways that people use categories, Rosch figured out that people use some simple categories, not as lists of membership criteria, but rather as if they extend from idealized mental representations (Rosch 1978). She used the word PROTOTYPE for these central members of mental categories, and PROTOTYPE EFFECT became the name for the pattern of behavior that Rosch had uncovered. The two most noteworthy features of this behavior and that are included under PROTOTYPE EFFECTS are (1) that categories seem to have both good and bad examples and (2) that they have fuzzy boundaries, such that membership for many instances and categories is uncertain or depends on the context.

The existence of prototype effects raises the specter of the brain and body within the process of categorical reference. They are evidence that the brain and the body bring idiosyncratic parts of themselves to the categorization process. If cognitive categories were lists of rules, there would be a simple and consistent procedural aspect to the way that we think about natural kinds, or categories in general, and the way we make categorical references. But prototype effects belie such austerity and suggest that the mind has a more directly representational way of working. A word that is often used in this context is EMBODIED, for it seems as if our cognitive apparatus embodies the very aspects of the world that become the subjects of thoughts and language.

But the implications of the overturning of the classical view of categories do not end at a new view of categories; they also bear on meaning and objectivity. From the finding that human conceptual categories strongly reflect the makings of the cognitive apparatus, comes the realization that thoughts and language about the real world are not objective in the way that they might be under a more classical, rule-based method. Under the classical view, the brain would be a sort of organizer of rules and symbols, and language would be an orderly conveyor of these rules and thus of information about the real world. Furthermore, the meaning of thoughts and statements would simply follow from the correspondence between those thoughts and statements and their real world counterparts. But this objectivist view of meaning does not hold if our categories do not mirror external reality. If, instead, the categorization process is also partly a function of something internal—the body and the mind—then the meaning of our thoughts and words regarding categories also contains something internal, and objectivity is at least partly sundered (Lakoff 1987).

Where does all this leave the pragmatic investigator? One might well have hoped for a simple correspondence between language and reality, perhaps fully embracing the classical view of categories, at least as they exist in language. Words for categories may not correspond to real entities, but we might have hoped that they still mirror the world in the classical way, and that their utility arises from a meticulous and orderly listing of properties within the brain. Certainly, many a pragmatic scientist would wish that we could take thoughts and words about the real world at face value and not have to consider their entanglement with the physiology of perception and thought. But the doubts that have been raised on the classical view of categories, on the general correspondence between language and reality, and on the objectivity of thought and language, are important and complicating news for any scientist.

Is it possible though, that where one model has been shed, another has been revealed? In rejecting models of the mind as a classical categorizer, have Rosch and her followers revealed a new improved model? As nice as this might be, it is not simply the case. The partial overturning of the classical view of categories has lead to a variety of theories on the way the mind categorizes. Lakoff devoted the heart of his book to a "theory" of what he referred to as idealized cognitive models. He was concerned that a basic prototype theory of categories—in which an object is categorized as an X, and not a Y, if it is more similar to X's best mental representation (i.e., prototype) than it is to Y's—is not really much of a theory. Unfortunately his alternative was no more than a listing of the various things a successful theory would have to explain. Conspicuously, it was lacking in any prospective predictive capacity and thus it was not testable (Green and Vervaeke 1997; Vervaeke and Green

1997). Rosch (1978), too, was skeptical that prototypes, or rather evidence that the mind makes judgments of prototypicality, could constitute a theory of categorization. A full theory must have some kind of description of what the prototype is and how it is built and how it is used by the mind to do all the things we use categories for. Today, however, a fairly literal prototype theory still persists and is the subject of ongoing psychological studies of categorization (Hampton 1995). Other related theories are also being investigated. A common theme among theories that draw on the discoveries of prototype effects is the idea of similarity. One class of ideas started by Medin and Schaeffer includes what are called EXEMPLAR THEORIES. They resemble literal prototype theories, for they hold that something is categorized as an X and not a Y if it is more similar to the other things that are already in X than it is to those things that are in Y (Medin and Schaeffer 1978; Nosofsky 1984; Smith and Medin 1981). Prototype and exemplar models of categorization both employ the idea that categorization behavior depends on the similarity between the item to be categorized and the category's mental representations. Nor have rule-based theories been entirely cast aside. Today the debate still plays out: whether the mind assigns categories by assessing similarity or whether it relies on rule-based criteria (Hahn and Chater 1998).

Perhaps the strongest critique against prototype or similarity-based theories is a question regarding the conceptual role of prototypes. A central focus for many who puzzle about cognition is the question of how the mind has concepts, and whether prototypes could be the basis for concepts. In this context, CONCEPT is roughly synonymous with CATEGORY, though concepts are generally taken to be highly combinatorial in a way that is not necessarily assumed for simple categories (Osherson and Smith 1981; Smith and Medin 1981). For example, the concept of a horned cow includes the concepts of horn and cow. The primary critique is that if simple concepts exist in the mind as prototypes then it is difficult to see how they could be used to generate more-complex combined concepts. Consider the example of the PET FISH category that was devised by Foder and Lepore (1996). Suppose, for the argument, that one's prototype of the PET FISH category is a goldfish, or something goldfish-like. This works for me, and probably for many people who grew up in places where goldfish are a common pet fish. But now consider what prototypes people might have for the FISH category and the PET category. For me, it seems like the FISH prototype is larger and sleeker and more silvery than a goldfish; and when I think of PET, I picture a puppy dog. The point is that my PET FISH prototype just does not seem to get anything from my prototypes for PET and FISH. But surely we must grant that the PET FISH concept does share something important with both the PET and FISH concepts. Indeed, whatever is shared is very important, for that sharing must be how complex concepts have meaning (Fodor 1998). And yet, it is not the prototypes that are shared. To put it another way, with a different example: it may well be the case that we have a sparrow-like prototype for birds and this may well play a role in deciding if something is, or is not, a bird; but the prototype itself is not an idea of what birds are. At least it is not the most important part of that idea, the part that permits one to hold complex ideas that include birds.

Foder, a philosopher who has written extensively on cognition, embraces the evidence, indeed the reality, of prototypes, but is also convinced that they are not the most interesting, or important, part of our concepts (Fodor 1998). But even if correct, there are many who still consider prototypes to be essential parts of our cognitive apparatus. Thus, for example, it may be that we do not need a similarity-based theory in order to explain all of our categories, and that the mind draws upon prototypes for many simple categories and is also capable of handling rule-based categories as necessary (Atran 1990, 55; Gentner and Medina 1998; Hahn and Chater 1998; Smith et al. 1998). Note that classical categories do emerge profusely in technical and scientific discourse where people often do rely upon narrow definitions (Smith et al. 1998; Ungerer and Schmid 1996, 40). In this context, at least three points deserve mention. The first is that we are capable of mentally processing categories in a way that at least seems entirely classical. Biologists do this all the time. They may have a sparrow-like prototype at the ready, but you can be sure that they regularly make use of rules (e.g., "birds have feathers") when they need to take care. Second, we regularly recognize and embrace the process of definition as one in which the criteria for membership in a category are itemized (though it is not always a process that works very well—chapter 3). Finally, language contains discrete symbols (words) and carries information in the sequence of words; it has a digital formalism that fits the classical model of categories better than the typological model. Under the typological model, a category is based on some central focus and instances can be close to that center or be less typical and farther from it. But to express the idea that "this item is an instance of category X," there is necessarily an all-or-none connotation. For the listener or reader, interpretation of such a sentence must begin with a capacity for handling classically categorical statements. But for adjectives and adverbs that can modify this connotation, the relationship entailed in all subject-predicate pairs must have this classically categorical aspect. These considerations suggest that prototypes, though they seem to lie at the root of many categories, are also overlaid by a capacity and a willingness to parse entities into discrete categories that are based on rules of membership.

Prototype effects seem to reflect basic features of the way that the mind contains and deploys information about the world, particularly for basic natural kinds, and for our purposes in this book, that is enough. It is enough for us to know that the process of using categories to make many decisions carries with it not a listing of properties (at the very least, not simply a listing of properties) but something more directly representational—an embodiment within ourselves of something that typifies a category. That embodiment or representation may not yet be well described by a theory, but whatever it is it gives rise to the effects, wherein many categories (many basic natural kinds certainly) seem to have good and bad examples and fuzzy boundaries.

Return now to the concern at the beginning of the chapter, that was over the quality of the relationship between language and reality. Under an objectivist view, which embraces the classical model that categories are bounded and organized within our minds by shared properties, the limits of language are mostly a matter of limits to the size of the mental database. Such a view does not lend itself easily to a question of whether language could somehow be biased with respect to some aspects of nature. But if our language about natural kinds also bears remnants of the way that natural kinds are represented in our minds, so as to generate prototype effects, then the question over quality seems much more reasonable. In short, there may be circumstances where our references to natural kinds are defective because they are encumbered by our cognitive apparatus. The word ENCUMBERED is appropriate, for as scientists we shall not give up the objectivist ideal. Even if we must acknowledge that the mechanisms that underlie our thoughts and language have been revealed as not perfectly suited to that objectivist ideal, we should still hope to identify such imperfections of language as hinder our understanding of nature.

This chapter's description of the uncovering of categories and of some claims against our objectivity might seem banal and a bit unworldly to the many psychologists and philosopher's who have grappled with categories in recent decades. However my purpose is not a review of that rich and complex literature, but rather to bring some of the lessons home to biologists. They are necessary if biologists are to appreciate our categories of organisms, as categories. Species are quintessential categories (though they may also be more than that—a topic for later chapters), and biologists should appreciate that categories exist first and foremost in the mind as organizational tools. Whatever method the mind uses to form and hold categories, it does so in ways that seem to draw upon idealized representatives of them.

The application of insights about categories, to species, may also be of interest to those in the cognitive sciences who are already quite familiar with categories. In the species problem, I think we find a very interesting application for the insights that have come from the study of categories. In fact, species (as categories) have been the subject of considerable research, as we shall see in the next chapter.

FIVE

TYPOLOGICAL THINKING ABOUT SPECIES

Now we are at one of those moments in an investigation when a seem-ingly tangential thread ends up leading to a useful insight. Meanders that return home with some effect provide not only the direct benefit of that discovery, but also enlarge the repertoire of ways to pursue fu-ture investigations. The meander of the previous chapter, a deliberately detached perspective that ran from the philosophical debates over real-ity and categories, to the psychologist's discoveries of prototype effects, is like this. The story, told so far, has reached only to the reality and the psychology of categories. It has not yet returned to the question of whether our language and thoughts are ill disposed for the under-standing of species, but it is close. This short chapter lays the two re-maining pieces of our wander, one that is historical and biological and another that is anthropological.

The first of these pieces is the 19th century scientific revolution in which our basic model of organismal kinds, and of organismal di-versity, underwent the great transformation from essentialism to evo-lutionary thinking (Mayr 1982). Prior to that transformation the predominant beliefs about kinds of organisms were dominated by Lin-naeus's typology, a well-codified version of Platonic and Aristotelian essentialism. Essentialism is the idea that a group has some sort of ideal, ethereal, representative standard and that each individual of the group is burdened by the degree to which it departs from that standard. But with the rise of geology (that revealed many fossils in rock strata) and advancing natural history, and the evolutionary thinkers of the 19th century, this view was overturned. Darwin delivered a one-two punch that pretty much obliterated essentialism. He showed that evolution oc-curred, and thus that there could not be kinds of organisms with fixed essences. He also showed, by his correct explanation of the mechanism of evolution, that the variation among individual organisms is the germ of variation among species. The complete idea, that variation among organisms is the crucial stuff of changing life and of life's progress, is devastating to essentialism. It says that variation among organisms of the same species, rather than being some sort of impure noise, is actually the mainstay of evolution, and thus of life. Today we well know that whatever real species are, out in the world, they do not have essences and they are not based on idealized representatives.

The second and final piece of the story, that brings our questioning of language back around to species, comes from the anthropological study of the ways that people in local, traditional, nonliterate societies think about and refer to different kinds of organisms. These studies are primarily interviews, where anthropologists have inquired of people about their names for the organisms that live near them. Multiple studies of this type have been conducted on diverse cultures, and it is possible to draw comparisons among them. The basic methodological idea is that by comparing multiple separate societies, with long independent cultural traditions, and that hopefully have not incorporated too much of the culture of the investigators, one can distinguish those features that are common to humanity (human nature, if you will) from those aspects that are more plastic (Brown 1991).

In order to describe these findings, it will help to first have some explanation of TAXA, which is the word for named categories of organisms. Many taxa are well known by their common names (e.g. MAMMAL), though biologists will often opt for the "scientific" name (e.g., *MAMMALIA*). Scientific taxa receive not only a name, but also a rank, such as Class (e.g., *Class Mammalia*) or Phylum, or Genus, or Species. These taxonomic ranks are ordered, by tradition, in a nested hierarchy, with SPECIES as the name of the rank that is basal to all others (i.e., species are the most numerous taxa, and they do not include other taxa) and KINGDOM as the name of the rank of the most inclusive taxa. All named species are taxa, and most formally named species also fit within many more inclusive taxa, one for each rank between Species and Kingdom. Finally, in the time since Darwin, the taxa of trained biologists have gradually and increasingly come to be used with the idea that they contain organisms that are evolutionarily related, with more relatedness among smaller taxa of lower ranks.

Taxa are natural kinds, categories, of organisms that people have devised to organize patterns of biological diversity. They can be studied in traditional biological ways (e.g., "What do the members of a taxon

have in common?"), but they can also be studied as part of human behavior. That is what some anthropologists have done, and from the
comparisons among their studies, it has been possible to identify commonalities. In particular, Berlin's (1992) synthesis of studies on multiple
isolated societies identified several common features of local taxonomic
systems. The primary findings are that traditional societies do have distinct names for kinds of organisms (i.e., they have taxa in a general
sense) and that a majority of these taxa tend to coincide with taxa that
are recognized by visiting "experts." Thus for example, among the Fore
People of the New Guinea highlands there are 182 basal taxa (i.e., taxa
that do not include other taxa) of vertebrates and nearly all of these
correspond directly to species taxa recognized by European systematists
(Diamond 1966). There is also a correspondence between the plant
taxa of visiting scientists and those of Tzeltal-speaking Mayan Indians
from Chiapas, Mexico, though it is not so strong as in the case of the
vertebrate taxa of the Fore. Only about one third of Tzeltal basal plant
taxa match up directly with species devised by visiting scientists (Berlin
et al. 1966), but of the others, there is still a very strong tendency for
taxa to correspond, though what is basal for Tzeltal speakers is often
not for visiting scientists, and vice versa (Berlin 1999). By themselves,
these observations are strong evidence that the taxonomic distinctions
drawn by professional biologists are not arbitrary, at least not to the extent that they vary freely from society to society (Diamond 1966). It is
evidence that many of the distinctions that we make also follow real
fissures in the biodiversity landscape.

A second major finding is that local taxonomies are arranged hierarchically, based on organismal similarities. Like the hierarchical taxonomies of scientists, local traditional taxonomies are nested in a way
that is roughly consistent with evolutionary history, though there tend
to be far fewer ranks in the hierarchy than found in those devised by
biologists.

Perhaps the most intriguing aspect of traditional classification systems is that they are based on biological affinity—similarity of appearance and lifestyle—and not on utilitarian issues. One could imagine
a classification scheme that includes a variety of practical criteria for
grouping (e.g., edible, near the home, dangerous) but that is not the
taxonomic system that arises in human societies. However, given that
these taxonomies are based on biological affinities, it is not surprising
that they have a nested hierarchical structure, for that clearly is a basic
feature of organismal diversity, species uncertainty notwithstanding (see
chapter 10).

Berlin describes other patterns that are found repeatedly in separate cultures, particularly with regard to the details of the hierarchy, on the number of ranks, and the number of taxa that tend to occur at each rank. But for our needs, the most important discovery to emerge from these anthropological comparisons does not concern the taxonomic systems, but rather the way that traditional peoples refer to kinds of organisms. The taxa of diverse traditional peoples exhibit strong prototype effects, just the same sorts of behaviors that were discovered and described by Rosch and other psychologists who have investigated how the human brain carries information on categories. These prototype effects are apparent as a major component of reference to organisms. No matter where a person lives, and no matter the organisms that they live with, that person's taxa are based at least partly on mental prototypes. Although manifest to some degree at all taxonomic ranks, prototype effects emerge most strongly in reference to organisms as members of the most basal taxonomic ranks. In these cases, though a taxon may be represented by many organisms, there is a clear tendency for some organisms to be recognized as prototypical for that taxon, whereas others are thought of as peripheral. In short, traditional folk taxonomies are deployed by native speakers in a typological manner, with references to kinds of organisms that are inspired partly by mental prototypes. Also, please note that the people in these traditional societies are not referring to organisms in an overtly typological way, nor indeed in any way different than what all people do. We all say things like "This tree is a maple," or "Lassie was a dog," and that is to use a language seemingly constituted of classical categories. Remember that prototype effects are revealed by the study of how individual persons identify things, and not by how they construct their sentences. They lay hidden to philosophers of categories because of the much more apparent manner that categories appear as part of communication, and that manner resembles classical Aristotelian categories.

Now recall the essentialism of Linnaeus, and of pre-Darwinian natural historians. As well reasoned as the arguments of these biologists were, their typological thinking did not just happen to be convergent with traditional folk taxonomies. Rather, the typological thinking associated with pre-Darwinian natural historians became integral to the taxonomies of those biologists because they drew on the local folk taxonomies of the cultures they lived in (Atran 1990). Nor was the inspiration of "scientific" taxonomies by folk taxonomies a bad thing, for this flow of knowledge lay at the heart of the renaissance of an empirical tradition in biology that superseded the remote intellectual approach of the Aristotelian tradition.

The punchline is that essentialist thinking about kinds of organisms is not a bygone fad killed off by evolutionists, nor is it a sideline curiosity teased out in specific contexts by psychologists and anthropologists. It was what people, all of us, tend to do (Berlin 1992; Medin and Atran 1999).

We also know with impressive certainty, for the story is well told and well reinforced by research on a daily basis, that it is an inaccurate way to think (Hull 1965a, 1965b). Whatever the correspondence between taxa and evolutionarily related groups of organisms, we know for sure that real evolutionarily related groups of organisms do not have essences or prototypes. Never is one organism a finer instance, or more representative, of its evolutionary group than any another, even if it may seem that way to us.

The Knowingly Biased Scientist

No part of this argument is directed against the seemingly reasonable ideas that typologically motivated references are efficient and justifiable in a variety of contexts. Indeed we know this to be the case even for biologists, as we need only recall the typologically dominated, and still quite successful, days of pre-evolutionary biology. But what we do have is strong evidence—from investigations in biology, anthropology, philosophy, and psychology—that the human mind is predisposed to refer to kinds of organisms in a manner that draws on mental structures that are inaccurate representations of the reality that causes those perceptions.

This insight places a biologist in an unusual situation. It is commonplace for an investigator, in any context, to be aware of internal constraints, of one's own strengths and weaknesses. But those kinds of observations are made in awareness of variation among people, or of variation within oneself over different times. But here I describe a failing that is probably common to all people; a failing found not in comparison to others, but in comparison to a truer knowledge. Of course, this story could not have been revealed, and nor could the truer knowledge have ever been obtained, were not people capable of seeing beyond their typological constraints.

Now recall the nature of the species problem that was described in chapter 1. It is not a conventional scientific problem, as we do not ask of what information we require to answer it. What it is, in its most pervasive manifestation, is a lack of successful communication and consensus. Biologists do not talk about species well. Does not the finding of a widespread, underlying bias, that misdirects our thoughts and discussion, fit with this social conundrum? The bias is evidently not so strong that most or all biologists cannot overcome it regularly in many particular situations. But as much as scientists may have broken away from typological thinking in some contexts, and as much as they have found ways to revise and improve upon folk taxonomies, the persistence of the species problem suggests that our typological tendencies have not been vanquished.

The possibility that the ways we devise and use categories are a cause of the species problem is not something biologists are justified in ignoring. The evidence of typological thinking, gained by anthropologists and psychologists, is considerable. But we still have little explanation of just how severe the bias might be, or precisely how it contributes to the species problem. Furthermore, after all these related questions are addressed, we must still ask how our categorical predilections might be overcome. All of these questions are more difficult than the one that has been answered.

This chapter concludes the first step of our critique of language and thought about species. We pause with the understanding that we do indeed have a predisposition to misunderstand species, at least on some basic mental level where natural kinds are processed. But before we can engage the next set of questions that follow from this understanding, we must return in the next chapter to the reality of biological diversity. So far it has been sufficient to know that typological thinking about real species is wrong. But we have just raised some more pressing questions about our typological tendencies, and these questions require a detailed view of biological diversity, one that is as free as possible from assumptions regarding the nature of species.

SIX

BIOLOGICAL DIVERSITY

We now return to more conventional biological turf to develop some appreciation for the way that biological diversity might really be, out in nature, apart and independent of our perception of it. With BIOLOGICAL DIVERSITY, I refer simply to variation, both phenotypic and genotypic, that occurs among organisms. At the outset of this inquiry, I must emphasize that our immediate goal is not a definition of SPECIES, and it is not a method for identifying species. Others have undertaken these tasks head-on and despite insights and successes on various fronts, the species problem still thrives. Instead we will proceed as we did with language. Just as we approached language from a distance, in order to avoid assumptions about the reality of references, we will also approach biological diversity from a distance. We do not need to assume that species exist—that the usual starting point of "What are species?" has any kind of straightforward answer—to try to understand patterns of biological diversity.

Some previous attacks on the species problem reveal fairly clearly the existential tangles that can arise when one begins with a strong presumption that species exist. The geneticist Alan Templeton once defined species as "the most inclusive population of organisms having the potential for phenotypic cohesion through intrinsic cohesion mechanisms" (Templeton 1989, 12). The statement is a tautology. If both instances of COHESION are replaced with something else—anything, even a variable like X—the statement still makes as much sense. Later, the tautology was patched up, and Templeton discussed various cohesive forces in detail (Templeton 1989, 1994). Michael Ghiselin, a biologist-turned-philosopher, bumped into a similar tautology when he considered defining species as the products of the process of speciation (Ghiselin 1997, 97). Both Templeton and Ghiselin are frank about the difficulty of finding strict definitions to describe species, and yet neither of them equivocates on the idea that species exist as real entities. Their method could be a page from the tautology cookbook: Take one strong species presumption and an equal measure of intangible reality, and squeeze out a definition.

The presumption that species do exist, and that we will try to avoid, actually has two parts. The first and most obvious is the presumption that organisms occur as part of distinct real entities that we might call species, and that every organism belongs to precisely one such real entity. The second common presumption is the idea that different species have something in common. It is this supposition of commonality among species that leads us to refer to a species category, the taxonomic rank that is named with SPECIES, and to suppose that this category is different from other categories that contain groups of organisms (e.g., other taxonomic ranks). These two parts of the species presumption correspond to the two common meanings of SPECIES: a real group of organisms (that meet the criteria of some species concept) and a taxonomic rank that contains many such groups.

One thing we cannot avoid, even if we are fastidious in avoiding the presumption that species are real entities, is references to kinds of organisms. Recall that language is full of statements that employ categories, and that kinds of organisms are quintessential categories. Suppose you take a ride on a horse. Is there some way to describe that event, or even notice it, without using the horse category or some other more inclusive taxon? Even a tortured description of what the horse looked like, and that avoided HORSE, would probably use the animal category. Thus we cannot avoid named categories of organisms, and a great many of these names are species names. But we can avoid assuming that a named category also refers to a real entity out in nature. We can avoid, for the sake of the discussion to follow, the assumption that all horses, or all the members of any species, collectively form a real entity. Just as we have words for thousands of natural kinds without assuming that all of the things in each of those kinds collectively form a real entity, so too can we have words for kinds of organisms without assuming that a kind of organism is also a real thing in nature. Recall the pragmatic, nominalist stance that was described in chapter 4. Here I am asking readers to begin a consideration of biological diversity by adopting that pragmatic stance, not just for most natural kinds, but specifically for kinds of organisms.

Avoiding the species presumption also does not mean that we must deny the marked structure that exists in the patterns of diversity among organisms. There are indeed reasons why we devise categories of organisms. The organisms that we label with HORSE really do resemble one another, and they really do not look much like most other organisms. The basic point, which is not in dispute, was put well by Dobzhansky "If we assemble as many individuals living at a given time as we can, we notice at once that the observed variation does not form any kind of continuous distribution. Instead, a multitude of separate, discrete, distributions are found. In other words, the living world is not a single array in which any two variants are connected by unbroken series of intergrades, but an array of more or less distinctly separate arrays, intermediates between which are absent or at least rare (1937, 4).

Note how this statement flows seamlessly from observations to reality. From "we notice" to "the living world is" it is a perfect example of the very common, and usually necessary, assumption of the omnipotence on the part of the observer. Generally, one cannot address reality without making this assumption to some degree. But Dobzhansky's statement is a useful example, for our purposes, and for making this distinction. Our starting point can be the same as Dobzhansky's, for we may take as correct, or given, both that people notice separate, discrete distributions within the patterns of variation, and that these patterns are real features of biological diversity. We can recognize this even if we avoid the species presumption and hold off from assuming that species are real, tangible entities.

Evolutionary Groups

Let us ask, in a simple and reduced way, how the patterns of strong similarity and dissimilarity that we find in biological diversity might have arisen. Most biologists who have studied biological diversity have focused on the diversity that is found among organisms. But for our purposes, it is better to cast the discussion primarily in the terms of molecular replicators, DNA molecules in particular. This reduction is motivated by our method which is to step back from the most pressing manifestations of our species uncertainty, and draw instead from the theory of life's origins as described in chapter 3. It is a theory of origins by the process of evolution and in its most basic form it does not even mention SPECIES or have any particular explanation for biological diversity. What the theory has, and what has characterized life on earth since its origins, are molecular replicators. These molecules leave descendant molecules with varying degrees of success, and that variation in replication success depends, in part, upon variation in the replicators. Capable replicators leave more descendants, and in a world of finite resources this success must ultimately come at the expense of the less-capable replicators.

In the theory of life's origins, organisms are but a modern evolutionary innovation; and even today, when organisms are the most tangible display of the diversity of life, we recognize that they are but the phenotypic manifestation of their DNA genomes. It is the DNA that carries the information for the organism and that is replicated, while the phenotype is the organism and is recreated each generation as a function of the genotype. Our focus will be on the transmission of the carrier of information, and much of this discussion should apply in principle to any informational replicator (Orgel 1992).

To help with the precision of what follows, I will use "DNA" to refer explicitly to a molecule that carries the information necessary for its own replication. A DNA may be physically connected to a longer contiguous stretch of DNA or it may correspond to a single chromosome, depending on the context. A DNA sequence is a particular order of the component nucleotides within a DNA and is not synonymous with a DNA. Thus, for example, a sample of two DNA molecules may include two DNA sequences or just one, if they are identical. The terminology also includes "DNAs" to refer to multiple pieces of homologous DNA. In this context, HOMOLOGOUS means that the different DNAs are related by common ancestry and thus share a history that can be represented by a branching tree diagram, sometimes called a gene tree. Figure 6.1 shows how an instance of DNA replication can be represented as a branching event. Figure 6.2 shows how a tree-shaped diagram can represent the key moments in history, the replication events, of a sample of homologous DNAs.

Consider a DNA that undergoes replication to form two daughter DNAs, and suppose that the replication depends upon both the DNA sequence and the local environmental resources. After replication, the fates of the daughter DNAs may be linked because they coexist under common circumstances and because they compete for the same pool of resources. If resources are limiting and competition occurs so that not all DNAs undergo replication, and if both daughter DNAs and all of their descendants are subject to the same circumstances (i.e., no mutational differences or geographic separation), then the long-term persistence of both groups of descendants is mutually exclusive. After some time, perhaps after many rounds of replication, one group of de-

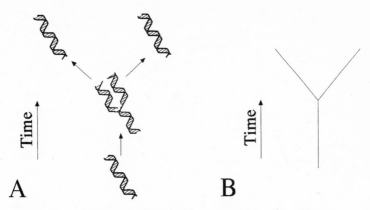

Fig. 6.1. (A) A double-helical DNA molecule before, during, and after replication. (B) The same replication event in a stylized fashion, using a straight line to represent the persistence of a DNA molecule through time and representing the replication event as a branch point.

scendants will have replaced the other, or both will have been replaced by the descendants of yet another DNA that also shares those circumstances. Furthermore, within any one group of descendants, there is a continuous turnover of patterns of ancestry. Consider the gene tree of ancestral relationships for a group of related DNAs, as in Fig. 6.2, and imagine the tips of the branches moving forward in time, as DNAs persist in time. Figure 6.3 shows the changing history of a small sample of DNAs at three time points. Among identical replicators under identical circumstances, some DNAs persist and replicate, and others perish. This random process, that is on average half replication and half death, leads to a shifting pattern of ancestry. As time moves forward, some lineages drop out, while others expand, and the nodes of ancestry that exist at one time may no longer be present at a future time. Indeed, this forward shift of ancestry will include occasional forward jumps for the bottom-most node representing the most recent common ancestor for an entire group of DNAs (Watterson 1982).

Now consider that most DNAs reside within organisms, and that the continuous random replacement of DNAs by the descendants of others is caused by a random birth and death process that happens within a group of organisms that share a finite set of resources. The random death of some organisms means that some gene tree tips do not persist, and the branches that lead to these tips disappear from the gene tree history that remains for those DNAs that do persist and repli-

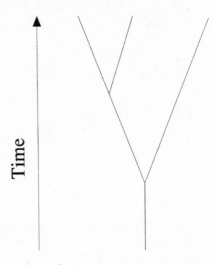

Fig. 6.2. A tree diagram representation of the history of a sample of three hypothetical DNAs. Key features include: the directionality of time, from the past to the present; branches; branch tips; and nodes, which are the junctions of branches. The three branch tips at the top of the figure refer to different pieces of DNA that exist at the present moment. The remainder of the diagram below the tips is a description of history. The tip of the branch at the base of the tree is taken to be undefined, as if the true history is not known beyond this point. Branches refer precisely to the persistence of a DNA sequence through time. This persistence means at times the physical persistence, but also includes numerous cases of replication when it is only the information in the sequence that persists. The nodes of the tree refer precisely to those cases of DNA replication in which both daughter DNA sequences that were the result of replication are ancestors of sequences that are represented as tips of branches.

cate (Fig. 6.3). Over time, the effect of this random population process is to cause most DNA lineages to be lost, while a single random DNA comes to be the ancestor of all DNAs in the population. Population geneticists call this random process GENETIC DRIFT, and it is often contrasted with nonrandom processes by which the DNAs of populations change, such as natural selection.

Two kinds of events can cause the descendants of two daughter DNAs to not be mutually exclusive. First, the daughters may differ because of mutation, and this may cause differences in the circumstances of replication. Individuals carrying the mutation may utilize resources

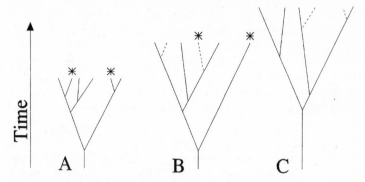

Fig. 6.3. A tree diagram representation of three successive times (left to right) in the history of a small group of DNAs. Asterisks (*) at branch tips at times (A) and (B) indicate sequences that did not persist to the next time period. Solid lines at times (B) and (C) indicate branches that were present at the previous time and still remain (and are now longer) in the tree because of the persistence of DNAs at the branch tips. Dotted lines at times (B) and (C) indicate new branches leading to DNAs that arose by replication since the previous time period. All solid lines at times (B) and (C) correspond to a line (solid or dotted) at the previous time, and all lines at times (A) and (B) (solid or dotted) that do not lead to an asterisk correspond to a solid line at the next time.

in a different way so that they do not compete directly with individuals that are not carrying the mutation. Second, one daughter DNA, and its respective descendants, may occur in a geographically distinct location from other DNAs. Under both mutation and geographic separation, the genetic drift experienced by the descendants of one DNA occurs partially independently of that experienced by the descendants of the other. To describe this in another way, the individual DNAs within a group compete more directly with one another, and are more likely to be replaced by the descendants of other DNAs within the same group than by the descendants of DNAs from the other group. In this way, both mutation and geographic separation can lead to multiple groups of DNAs that are not mutually exclusive.

This simple model of replication that leads to multiple groups of DNAs that are not mutually exclusive has three components: a DNA with a sequence that causes replication; the possibility of mutations; and some kind of environmental structure such that the pool of resources used by one group of DNAs need not completely overlap those of another group of DNAs.

The model is also an approximation for cases when DNAs are drawn from multicellular organisms. In these cases, the large majority of DNA replication events have happened during the growth of organisms, and only a few actually occurred in the cells that gave rise to new organisms (e.g., in the formation of sperm, eggs, pollen, ovules, spores). But a gene tree perspective can still be a good approximation. If a sample of organisms and of DNAs is considered such that not more than one DNA comes from each organism, then all the basic components of the simple gene tree model still hold. In this case the nodes of DNA coancestry that are on the tree will represent DNA replication events that happened during the making of cells that gave rise to new organisms.

The term EVOLUTIONARY GROUP will be used, both for a group of DNA replicators as described above, and for a group of organisms generated by those replicators. This kind of group is evolutionary because its members share a history of common descent, and because they all compete for common resources, but most importantly because they share in carrying beneficial mutations—indeed the group may be created by a beneficial mutation in a DNA that causes it, and its descendants, to partake of resources in a different way or in a different place. This kind of group would also meet the criteria of individuality and of existence. Existence is an attribute of entities, which are things that have some tangibility and are localized in space and time (chapter 4). Though consisting of multiple separate replicators, an evolutionary group is an entity because its components are interacting and competing, and because the location of its members are constrained in space and time (Ghiselin 1966, 1974; Hull 1976, 1978). People are quite familiar with this kind of highly dynamic, non-corporeal mode of existence. Consider a family, or a colony, or an ecosystem, or a committee, or a conspiracy, or a tournament, or any of a multitude of other things with which we are familiar and in which the components move around. Surely we consider many such things to be real. Existence is not undone just because the parts of a thing move around and are themselves entities.

This simplistic view of an evolutionary group, whose members share in competition, genetic drift, and adaptation, connects with a wide variety of views that others have expressed about species. Within an evolutionary group an individual organism can be physically replaced, and their function in the environment can be replaced, by the descendants of other organisms that are within the group. The organisms within the group compete, so that the survival and reproductive success

BIOLOGICAL DIVERSITY   75

of one organism has an affect on the survival and reproductive success of other organisms. Fisher wrote of species as entities enjoined by processes of competition and adaptation (1958, 135–139). This type of competition and potential for replacement has also been described as demographic exchangeability, a key component of the cohesion species concept (Templeton 1989). Such competition also occurs when organisms share the same niche, in an ecological sense (Hutchinson 1958). The idea that organisms within a species share a niche is also a component of several other species concepts, including the evolutionary species (Simpson 1961), as well as one formulation of the biological species concept (Mayr 1982, 273), and it is contained within the ecological species concept (Van Valen 1976). A similar view, emphasizing the competition that occurs when organisms closely share resources, has been described by Ghiselin (1974) in an exposition on species as individuals (as opposed to categories). Ghiselin proposed that SPECIES be defined as "the most extensive units in the natural economy such that reproductive competition occurs among their parts." A common thread among most species concepts, and contained within the idea of an evolutionary group, is that the species is the place where the processes of evolution happen.

Recombination

When a DNA molecule, or a part of it, joins with another DNA molecule, then recombination has occurred. It can happen in a variety of ways, and the DNA genomes of different kinds of organisms vary widely in just how frequently it occurs. Some DNAs seem to never do it, while others do it every time they replicate. In a gene tree view of the history of a sample of DNAs, recombination is any process that causes different portions along the sequence of a set of DNAs to have different gene tree histories. In the absence of recombination, the gene tree history of a sample of DNAs is the same for all parts of the sequence. With recombination, one portion of the DNA will have a different gene tree history than will another. Furthermore, it is possible that the speed of genetic drift for one stretch of sequence, in a sample of DNAs, may be faster than it is in another portion. If there is a high recombination rate, then a large sample of DNAs will have a history of many different gene trees, perhaps as many as there are nucleotide base pairs in the sequence.

The causes and effects of recombination will be more fully explored in the next chapter. But there is one context where we can consider recombination in fairly short order, and where we can appreciate that it does not greatly alter our view of evolutionary groups. Consider the simple model of an evolutionary group that has been described, within which there is competition and shared adaptations, and turnover of ancestry. If we allow that recombination also occurs within such a group, then very little of our basic view is altered. Thus, for organisms that are obligatorily recombinogenic—which must do it in order to reproduce, such as most eukaryotes (organisms whose cells have a nucleus)—an evolutionary group is very similar to what Dobzhansky called a Mendelian population (named after Gregor Mendel): "A Mendelian population is a reproductive community of sexual and cross-fertilizing individuals which share in a common gene pool. . . . The smallest Mendelian populations are panmictic units (Wright 1943), which are groups of individuals any two of which have equal probability of mating and producing offspring" (Dobzhansky 1950).

Thus by definition, organisms within a Mendelian population share in a probabilistic process of reproduction, and all pairs of organisms are equally subject to reproductive failure and equally likely to reproduce. Within a Mendelian population, each generation occurs with some distribution of reproductive success among the component organisms. The shape of this distribution may vary across generations, but at any point in time the particular pattern of reproduction is a major determinant of the gene tree for all portions of the genome. A sample of DNAs for a short region of the genome will have a particular history, whereas a different genomic region will have a different history; yet all of these histories must run through the same historical procession of organisms, with a different group of reproductives each generation. Thus a Mendelian population carries genomes with numerous gene trees that were all shaped by a common birth and death process.

When a beneficial mutation arises within a Mendelian population it causes a rapid turnover of ancestry for a portion of the DNA, but just that portion that is tightly linked to the mutation. A Mendelian population can undergo adaptation via fixation of a beneficial mutation while leaving most of the genome unaffected by the selective process. Not so an evolutionary group that does not engage in recombination, where a beneficial mutation quickly and greatly shortens the entire gene tree history for all the DNAs, and for each along its entire length. But in both contexts, with and without recombination, the basic competitive process occurs. What is different within Mendelian populations is that some portions of the genome have faster turnover than others.

Though this is due to differences in natural selection for different portions of the genome that can have different gene tree histories by virtue of recombination, it is just as if some genes are undergoing fast genetic drift, while others are undergoing slow genetic drift. Indeed, at the level of DNA where there is linkage, natural selection on functional DNA sequence variation contributes to the genetic drift that occurs among linked sequences. For the DNAs of organisms with recombination, the acceleration of genetic drift by natural selection depends on the degree of linkage, the number of sites of functional variation, and the strength of natural selection on the functional variation (Felsenstein 1974; Hill and Robertson 1966).

As much as recombination is a major evolutionary force, it does not alter the basic picture of the evolutionary groups that have been described here. These are competitive groups that share in a process of genetic drift and that share in adaptations, and none of these features are removed if recombination occurs among the members of a group.

The Causes Of Multiple Evolutionary Groups

As mentioned above, there are two kinds of events that can cause a single group to become two. First is physical distance or the emergence of a physical barrier between DNAs (or their organisms) so that they do not draw from the same pool of resources. This geographic barrier to drift may be reversed if organisms are mobile or if geography changes. Second is the appearance of a mutation that changes the sequence of a DNA so that an organism and its descendants undergo genetic drift separately from other organisms not carrying the mutation. This kind of origination of an evolutionary group can be reversed only if all of the descendants of the organism fail to reproduce so that all copies of the new sequence ceased to exist. Back mutation could not undo the event unless all descendant copies of the mutant DNA underwent back mutation.

The effects of geographic barriers and environmental heterogeneity on genetic drift do not change as a function of recombination. If geography constrains the replacement of some individuals by the descendants of others, then it also constrains the process of recombination between some pairs of individuals. This constraint on genetic drift will occur for all portions of the genome regardless of recombination.

The kinds of mutations that can create evolutionary groups do differ somewhat with recombination. In the absence of recombination, a new advantageous mutation causes an organism and its descendants to undergo a different pattern of genetic drift from those organisms not carrying the advantageous mutation. If this does not cause the extinction of those DNAs that do not carry the mutation, then there are two groups where once there was one. But when there is frequent recombination, favorable mutations do not contribute directly to the multiplication of groups. Regions of the genome under tight linkage to the site of the mutation will experience accelerated genetic drift and have a shortened gene tree history (Kaplan et al. 1989; Maynard Smith and Haigh 1974) while the gene trees of unlinked portions of the genome will not be affected. In short, recombination can prevent favorable mutations from causing one group to split into two, because organisms that lack the beneficial mutation are not excluded from the birth and death process that occurs among organisms that carry the mutation. However, there is a class of mutation that, in obligatorily recombinogenic organisms, can contribute to the formation of new groups. These are mutations that cause recombination either not to occur between some individuals or cause the results of recombination to fail to reproduce. Included within this class of events are genomic changes that shift the mode of reproduction or the ploidy level (the number of copies of each individual DNA, or chromosome) of the genome. For example, interspecific hybrids of recombinogenic organisms may be polyploid (having more than two copies of all chromosomes) or parthenogenetic (capable of reproducing without engaging in recombination with the DNAs of another organism). In either case, the progeny of the hybrids can no longer exchange genes with members of the original group and do not share in a common process of genetic drift with either original group.

This view of evolutionary groups is simple as it follows fairly directly from the idea of molecular replicators in a finite environment. This reduced view also helps with an appreciation of the causes of biological diversity, and as suggested above, there are commonalties between these ideas, on the causes of diversity, and those contained within well-known species concepts. Having posed the discussion in terms of gene tree histories and genetic drift, we find that mutations and barriers to the movement of DNAs appear to be the only possible causes of biological diversity. Similarly the irreversible aspect of mutation, as a cause of diversity, contrasts with the reversible effect of barriers, and this distinction follows directly from the genetic drift perspective. However, these mechanisms are not novel ideas, but rather a simple version of the causes of speciation that have been discussed in other contexts. In particular, geographic isolation and the evolution of barriers to recombination form the most widely discussed model of speciation (Dobzhansky 1937; Mayr 1942). Also, a similar depiction of the impact of different kinds of mutations, and their shifting effect as a function of recombination, is contained within the cohesion species concept (Templeton 1989).

Readers may notice that all of this discussion pertains to evolutionary groups, and that I have avoided equating them with species. The main reason for this is that SPECIES is a term laden with ideas about distinctions between kinds of organisms. As simple and reasonable as the theory of evolutionary groups might seem, it flows entirely from the idea of molecular replicators and draws not at all from information on variation among real world organisms or DNAs. Nor does the idea of evolutionary groups help, at least in any way that is nearly so simple as its genesis, with thinking about how to recognize kinds of organisms or how to tell them apart from one another. What the idea of evolutionary groups does connect with, much better than it does with SPECIES, is several aspects of the species problem, particularly those many cases where species are difficult to identify.

Boundaries

If one considers the existence of things—entities, generally—and also considers language about the real world, then there are various ways to appreciate that the things in the world are often not as simple as they appear when described with words and sentences. We saw in previous chapters how the boundaries of entities are often diffuse, or indistinct. As mentioned in chapter 4, even that most tangible of entities, the earth, has an atmosphere and a Van Allen Belt that extend into space, increasingly diluted with distance, in a manner that does not form a sharp boundary for our planet. So too are the boundaries of the planet category difficult, in a different way from those for the earth, as there is necessarily an arbitrary component to deciding how big or how small an object orbiting a star must be to be called a planet. It is conceivable that arguments along these lines could be generated against the use of a name or a category for any and every material portion of the universe. It is not a small thought for a scientist concerned with the application of language to reality. We shall not embrace it entirely, but we can use it to help address a common uncertainty that comes up when an investigator is faced with an uncertain boundary, be it over a species or something else. When an investigator says a boundary is unclear, the statement could have two quite different existential meanings: (1) It might be that the boundary is truly vague, such that no sharp line exists to be observed, no matter how closely one might look; or (2) it might be that the boundary is only apparently vague, and might be revealed as fine and sharp upon a closer look. In principle the investigator cannot know the difference until such time as they are satisfied that they have found a sharp boundary. The example of the unbounded earth, and of the many other things that are truly fuzzy, reminds us that point (1) is an entirely plausible reality—there may be no sharp boundary. Let us bring this point home to the problem of a single evolutionary group that is in the process of becoming two. A biologist considering such a group might feel they must decide whether the group is one thing or two. In other words, they may think they have a counting problem. But if they do indeed treat the judgment as a counting problem, then this is tantamount to assuming that there is a true and sharp boundary to be found—they assume that meaning (2), above, applies. The alternative is to allow the possibility that the ambiguity is real, but in order to do so they must give up on counting.

An evolutionary group need not have sharp boundaries, in the sense of meaning (1) above (i.e., no sharp boundary exists). What were described as group criteria—shared genetic drift, competition, and adaptation—need not generate a sharp partition. The processes of genetic drift and competition and adaptation may all be occurring, but not necessarily in ways that they are all equally shared by a distinct group of organisms. In such cases, GROUP would be misleadingly distinct.

Competition is probably not a process that occurs equally and simultaneously among all pairs of a set of replicators, whatever their relatedness or geography. It will change with the density of replicators, the abundance of resources and the weather, to name some major variables. Adaptation as well, even in the most simple of situations wherein DNAs that carry a particular mutation are increasing in numbers, can be expected to be highly variable depending on all those same things that competition depends upon. Consider too the simple effects of geography, such as envisioned by Wright in the isolation by distance model (1943) (see also chapter 7). Under this model (which probably applies to most organisms to some degree) organisms of limited mobility are spread across a landscape, and the times of possible coancestry for a given pair of DNAs are proportional to the physical distance between the members of that pair. Under this scenario the pattern of genetic drift, as well as the pattern of genetic variation, among organisms may not be structured but may follow a continuous pattern over some environmental landscape.

Not all biologists are uncomfortable with real imperfect boundaries between kinds of organisms. One of the main reasons for this is the existence of theories on the gradual origin of species. Any theory where one thing gradually becomes two, necessarily admits the existence of real partial boundaries, of circumstances where counts cannot be made. Perhaps the best examples arise in consideration of the Biological Species Concept, which will be discussed in the next chapter.

Hierarchies

Another pattern that can be expected to arise within evolutionary groups is that of nested and hierarchical patterns of structure. Genetic drift, competition, and adaptation may be shared over a local scale, perhaps associated with a very small number of replicators. These processes may also be shared to a lesser degree over a broader circle, and still less so over a yet broader circle, and so on. Furthermore, any larger area of limited structure might contain multiple smaller regions of more structure. This nesting pattern was envisioned by Dobzhansky, who wrote of Mendelian populations nested within one another, each level associated with a particular level of mixing that is less than the smaller Mendelian populations that are contained within it (1950).

A hierarchical, nested structure would mean that evolutionary groups would seem to be apparent and numerous regardless of the scale of assessment. For example, a small sample of individuals spread over a wide area would show the effects of competition, genetic drift and adaptation that occurred at that scale, and it would miss the smaller better-defined groups that exist on more local scales within the larger group. A more-localized sample would also show evidence of evolutionary groups, but in this case reflecting the forces of competition, genetic drift, and adaptation that occurred on that more local scale. A large sample spread across both local and wide areas would show this hierarchical aspect, but unless every DNA of the large evolutionary group were included in the study, the finest limits of the evolutionary groups might still not be observed. The hierarchical pattern gives rise to an appearance of structure that is much the same regardless of the scales of observation. From a distance, or with a dispersed sample, one observes the larger structures, whereas from a short distance with a localized sample, one observed the smaller structures—but the essential features of the structures at different scales are the same.

FRACTAL is the famous and enduringly trendy name, devised by Mandelbrot (1977) for things that exhibit self-similarity across different scales of observation. Fractals are especially accessible as mathematical, graphical things, particularly the beautiful fractal pictures that are not difficult to generate with a computer. The fame of fractals comes primarily from two things: the very rich, apparently complex, patterns that arise from very simple mathematical models; and that a great many aspects of nature appear to resemble these patterns. Scores of mathematical fractals have been described, and one large category includes branching processes. The gene trees within evolutionary groups are a branching process, as are river systems, blood circulatory systems, and the growth form of many plants. Any particular branching process, like any fractal, seems to have the same design whether we are looking at the entire pattern or just a small portion of it. When we step back and look at the entire pattern of a branching fractal, we see the connections among large branches; but when we move closer we can no longer see these branches, but instead see the connections among smaller branches.

If evolutionary groups arise in the midst of a fractal landscape of other evolutionary groups, then we can anticipate some difficulties of measurement. The foremost of these is the inherent uncountability of structures within a fractal. Imagine a mountainous landscape with multiple peaks, on each of which are various bumps and rocks, and the rocks themselves are irregular and have more bumps. Also between the mountains are valleys, and within the valleys are hollows and gullies and holes of varying sizes. Much of Scotland has this sort of landscape, and a popular pastime in the highlands is to climb all of the Munros, the mountain peaks over 3000 feet in height, originally described and compiled by Sir Hugh Munro in 1891. According to the latest revised count, Scotland has 284 Munros. But, of course, it is not possible to count all peaks over any particular height, not without arbitrary rules to define peaks. A mountain might seem plain enough from a distance, but the closer one gets the less apparent are the big mountains, and the more one perceives their rough surface. Any one Munro will have rocks and irregularities on its surface and at its top. Is each bump on the peak to be counted as another Munro? The counting of peaks absolutely requires a set of rules—arbitrary rules—about how far apart two peaks must be and by what sort of chasm they must span, before

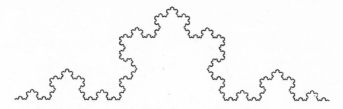

Fig. 6.4. A portion of the Snowflake Curve of von Koch (O'Connor and Robertson 2000).

they can be counted as two peaks. Figure 6.4 shows a well-known lin-ear fractal, called a Kock curve. Like our example of the mountains of Scotland, one could not count the bumps in the Kock curve. As a mathematical entity, the Kock curve has an infinite number of bumps. A fundamental feature of fractals is that they cannot be quantified by counting their parts. But there are ways that the structure of a fractal can be quantified. The Kock curve does have a finite length, calculable as a limit, and Scotland has a finite area in the same way. Also, Mandel-brot devised an index that reflects the degree to which a fractal fills the dimensions within which it resides.

The uncountability of structures within fractals is not limited to those that resemble actual landscapes. Figure 6.5 shows two plain branching fractals, both drawn by short computer programs. The one on the left somewhat resembles a tree branch, while the one on the right is a very regular iteration of the forked structures like that in Fig. 6.1. Though far more regular than real-world branching fractals that might arise (such as river systems or systems of blood vessels or evolu-tionary trees), it shares with these an uncountable number of parts. As mathematical entities, the fractals represented in Fig. 6.1 can have an infinite number of branches. In the real world, this limit will not be reached. A river system goes from rivers to streams to creeks and on down to the tiniest rivulets. A system of blood vessels has capillaries at the smallest scale. So too must a gene tree have individual DNAs at the very tips of the tree, and these must be finite in number. Thus one could conceivably count the entities at the finest of scales of real world things that have a branching fractal structure. But if one would like to count larger branches, then there may be no objective criteria by which to decide whether a branch is the correct size to count.

Evolutionary groups do not exist literally as gene trees, even though gene trees may describe their histories, and they are not branching frac-tals. They are competitive groups of related DNAs and of organisms

Fig. 6.5. Two branching fractals.

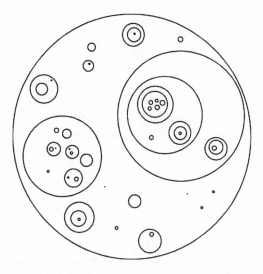

Fig. 6.6. A circle represents the border around a group of DNAs or organisms (not shown) that experience shared competition and genetic drift. Smaller circles, within larger ones, encompass smaller groups wherein competition and genetic drift are shared more strongly.

that carry those DNAs. They are related because their proximity and competition causes them to share in genetic drift and to share in the adaptive fixation of beneficial mutations. But the degree to which individual organisms interact with others, and share in the processes that cause an evolutionary group, will vary with geography and local circumstances. Another way to visualize the kinds of fractal hierarchical structure that might emerge within an evolutionary group is shown in Fig. 6.6, which is merely a drawing of circles nested within other circles in an approximately random fashion. The idea is to consider each circle as circumscribing a geographic area within which those processes that generate evolutionary groups are shared to a particular degree, with different degrees identified by different circles. The point is not circles, per se, but the fractal structure of nested levels of the magnitude of a process that is spread across a geographic area, just as one finds fractal structure in the pattern of bounded regions within a topological map.

One could, of course, count parts of fractals, just as Munros are counted, with the aid of a set of arbitrary guidelines. Such guidelines might not even seem arbitrary. People are not omnipotent observers across all scales, but rather fall in a narrow-size range. Even without the idea that Munros are higher than 3000 feet, it is not difficult to imagine that people would see mountains as countable and that they would agree, roughly, on the number of peaks. Their approximate agreement would arise from shared culture, and the flow of ideas about mountains within that culture, and from shared scales of observation. But imagine the counts made by a much smaller observer. Would not a mouse perceive as many peaks and as much rough geography within an acre, as people perceive in all of Scotland? Finally, consider the task of a scientist, or any observer who might wish for an understanding of nature that is not dependent on the scale of the observer. For any investigator who wishes to not be bound by their own scale of observation, the counts of structures within a fractal topography are quite inadequate.

Under the theory that has been devised here, evolutionary groups clearly exhibit a potential for fractal structure—evolutionary groups within evolutionary groups—and thus of uncountability because of that structure. The theory also corresponds well to Dobzhansky's view of Mendelian populations. In this light consider some common practices, among biologists, of ways of using the SPECIES category. One common view is that the SPECIES category should hold only basal taxa (i.e., individual species taxa should not include other taxa). Another common, and closely related, view is that species are terminal taxa, meaning that they lie at the tips of evolutionary trees that are used to connect species (which are different from gene trees: chapter 10). But if the hierarchical patterns of biodiversity are actually fractal in nature, then there may not exist patterns that could be used to devise a truly basal taxon, or a truly terminal tree tip. If biodiversity is fractal, then we would expect that one could always find a finer pattern within a pattern, a smaller group within a group, except in the not very useful limit wherein basal taxa include just individual organisms.

A Simple Theory of Diversity

This chapter has begun a simple theory by building on basic ideas about replicators under limiting circumstances. The theory is still missing some things (to be considered in the next chapter) but already we can make use of the idea of evolutionary groups and of their similarities to species. Evolutionary groups can be contemplated at either the level of the genes (the DNA) or at the level of the organismal phenotypes. Either way, they are created by DNA replication and competition, two factors that lead in turn to the evolutionary forces of natural selection and genetic drift and ultimately to the fission of evolutionary groups. The similarities to some common ideas about species are readily apparent on two levels. First, if our simple theory is fitting, then evolutionary groups exist—as real dynamic entities consisting of multiple interacting, but unconnected, parts. This kind of complicated entity is just what many biologists envision when they think of a species existing in nature. They think of an entity consisting of multiple organisms enjoined by their sharing in forces of evolution. Some species concepts say this explicitly, while some others imply it (though some others share nothing with these points). Alan Templeton's cohesion species concept is the one that most thoroughly embraces the ideas of shared competition, drift, and adaptation (Templeton 1989). Second, the idea of evolutionary groups show us something about the species problem. What we can see readily with hypothetical evolutionary groups are the ways those groups may not be distinct. Because of hierarchical structure within them, and partial boundaries between them, we would expect evolutionary groups to sometimes be uncountable. We see revealed in the theory of evolutionary groups, a parallel to what many biologists have found to be so frustrating about species, and that is difficulty in counting.

SEVEN

RECOMBINATION AND
BIOLOGICAL SPECIES

Most DNA molecules that are the genomes of organisms include genes that permit or encourage the breaking and joining of the DNA with other DNA molecules. In other words, most genomes include genes that cause recombination. This process of recombination is often highly choreographed and occurs regularly at specific stages of an organism's life cycle. For many eukaryote genomes it occurs at the life cycle stage of transition from diploid cells (with two copies of every chromosome) to haploid cells (with just a single copy of each chromosome). This process is called meiosis, and when recombination occurs during meiosis it involves two similar genome copies that had previously come together earlier in the life cycle at the time of transition from haploid to diploid cells (i.e., syngamy or fertilization). Many genomes have also evolved multiple disjunct DNA molecules (chromosomes) among which there is effectively free recombination at the time of meiosis. For the genomes of prokaryotes there is no meiosis and syngamy, yet the breaking and joining of DNAs may still occur with high frequency (Lenski 1993). Whether or not the shuffling of DNA segments occurs in the course of reproduction or at some other life stage and whether or not it proceeds by swapping portions of DNA molecules or by the incorporation of one DNA by another, recombination leads to a new genome that is an admixture of two.

The Evolution of Recombination

If we continue with the basic method of this book, to approach the topics of biological diversity and of language about that diversity from a distant and reduced perspective, then the next step is to extend our appreciation of the theory of life to include recombination. The evolution of, and evolutionary maintenance of, recombination is a large and complicated topic, about which much has been written from a variety of perspectives. Fortunately, our basic approach that extends from the simple theory of replicators leads to a fairly straightforward explanation of why recombination evolved.

Given the prevalence of systems of DNA exchange among prokaryotes and the fact that some basic molecular components are shared by prokaryotes and eukaryotes (Gupta et al. 1997; Shinohara et al. 1993; Sung 1994), recombination was probably an adaptation that arose (at least once, but maybe many times) early in the evolution of life (Margulis and Sagan 1986). Thus, despite the existence of many models on the evolution of recombination it is perhaps sufficient for the present synthetic purpose to explore those pressures that are expected under the most simplistic models of replicators, and that may have played a role early in the origins of life (and since then as well).

In fact there are some simple models that predict the existence of an evolutionary pressure for recombination, under which mutations that lead to recombination, or increase its frequency, are expected to be favored by natural selection. These models follow from consideration of the fact that mutations occur and cause some DNAs to be better replicators than others. The key point is that the persistence of a portion of a DNA depends very strongly on the replicating capacity of the rest of the DNA molecule to which it is attached. If the longer DNA is a good replicator, then there is a good chance that it will leave descendants, in which case the long DNA and all its portions share in this success. Conversely, if a portion of DNA is linked to others that collectively act as a poor replicator, then they will all perish together.

Now consider a portion of DNA that somehow, through the expression of genes it carries, causes recombination or acts to elevate levels of recombination. In the absence of any other beneficial mutations, these effects on recombination may increase the likelihood that the DNA persists and leaves descendants. This is so simply because recombination may cause a change in the quality of linked DNA, from linkage to a poor replicator to linkage to a good replicator. Of course, the reverse is true as well, as blind recombination may move a DNA into linkage with a poor replicator. But contrast two situations, with and without recombination. In the absence of recombination, a portion of DNA and all its descendants are linked to those of another particular DNA. In essence, a DNA has but one shot at success, and that shot depends strongly on the particular DNA it is linked to. Without recombination, the long term success or failure of a DNA is dependent on one particular linkage relationship. Furthermore, since most mutations are harmful and since only a few DNAs leave descendant over the long term, the fate of the DNA is probably negative. But with recombination a DNA can leave descendants that are linked to different DNAs. And if there is a chance for multiple different linkage relationships, then there is also a greater chance that at least one of those linkage relationships will form a good replicator and have good prospects for persistence. In short, a portion of DNA that promotes recombination will contribute to variation in linkage relationships among descendant copies of that DNA, and thus will have a higher opportunity for being favored by natural selection (Felsenstein 1974).

This narrative is simplistic, but it follows a more rigorous quantitative model. Otto and Barton (1997) modeled the evolutionary fate of a mutation that is a neutral modifier of recombination and that occurs in a finite population of chromosomes that are segregating multiple selected mutations. They showed that a mutation that has no other effect than to increase recombination will have an increased chance of fixation, relative to a neutral mutation that does not alter recombination. Indeed, from the perspective of the mutation that increases recombination, that effect conveys a selective benefit, with a contribution to the probability of becoming fixed in the population that is indistinguishable from that for a gene with a mutation that is beneficial in some more direct way.

Otto and Barton's results were limited to neutral modifiers of recombination, and in general the effect they describe is weak. But a more general model, in which the recombination modifier also has other effects on the replicating capacity of a DNA (i.e., it is pleiotropic), predicts a considerably larger effect. Mutations that are beneficial, and also raise the recombination rate, are more likely to leave large numbers of descendants than are mutations with identical effects on replication, but lack the effect on recombination (Hey 1998). Similarly harmful mutations that *reduce* recombination are more likely to leave large numbers of descendants, than are similar harmful mutations without the recombination effect. However such negative mutations are very unlikely to leave large numbers of descendants regardless of the recombination rate.

A related class of simple models, that favor recombination when there are multiple mutations, supposes that most mutations are deleterious and that they interact synergistically in their negative effects (i.e., each additional mutation carried by a genome causes a larger negative effect than the previous mutation). In the absence of recombination, this kind of synergism causes just those DNAs carrying high numbers of mutations to be removed by natural selection. But the flip side of the synergism of mutation effects is low variation in replicating capacity among DNAs with low and intermediate numbers of mutations. Once natural selection has removed those DNAs with high numbers of mutations, there is little scope for the removal of additional mutations because all the DNAs that are left have the same replicating capacity. However, if recombination is added, then the shuffling among DNAs tends to restore variation in the number of mutations. Most importantly recombination acts randomly, but steadily, to reconstruct genotypes with high numbers of mutations by bringing together portions of other genomes that overall do not have high numbers of mutations. The immediate result is that natural selection acts more effectively, and the longer-term result is that the mean fitness of the population is higher when recombination is occurring (Kimura and Maruyama 1966; Kondrashov 1984, 1988). The crux of this model is the synergistic affect of mutations, and it can be generalized to a variety of fitness schemes, with similar predictions regarding recombination (Barton 1995).

These are far from the only models that have been developed and that can predict the evolution of recombination (Kondrashov 1993). Nor do these models address the complexity of recombination as it occurs in many organisms, particularly the complex aspects of sexual life cycles that are associated with recombination (syngamy and meiosis) and that are so common among eukaryotes. In general the difficulty is not devising models that predict the evolution of recombination and sex, but rather finding ways to distinguish them experimentally (Barton and Charlesworth 1998).

The Limits of Recombination

But do we see, or should we expect, recombination among any pair of genomes? In fact people have long recognized a nearly ubiquitous manifestation of diversity among eukaryotes, that the more dissimilar are two organisms (or two genomes) the less likely they are to engage in recombination. At the organismal and behavioral level, those organisms that invest in mating behavior are often quite selective about their mates. At the level of zygote formation, two gametes either reject one another or the zygote dies if the gametes are not compatible. These words are just one way, that does not rely on SPECIES, for describing points that often arise in discussion of species. Those points are that gametes can only join if they are from the same species (under the Biological Species Concept, see below), and that organisms are more similar to those of the same species than to those from other species (similarity is either a de facto component or a prediction of all species concepts).

Prokaryotes also reveal strong relationships between recombination and DNA similarity. Though non-meiotic, prokaryotes exhibit recombination in a wide variety of contexts. Perhaps the most direct of these is the capacity to take up, and incorporate into the genome, DNA that lies outside the cell. But the rate at which this occurs is a strong declining function of the similarity of the imported DNA fragment, and the sequence in the genome. (Cohan 1994; Zawadzki et al. 1995).

To better grasp that recombination should decline with dissimilarity, return to the point mentioned above, that the success of a portion of a replicator depends on the success of the replicator as a whole. Whether or not recombination happens, the parts of the genome must work together to make a phenotype and thus to carry out replication. The more functional integration that exists across the genome, the less likely that a portion of that DNA could be expected to gain prospects by leaving and joining another. The basic idea here is that the more functional interaction that exists across parts of the genome, the more likely it is that a reducer of recombination will be favored. The mathematical theory for the reduction of recombination is actually most well developed for the case when multiple polymorphisms are segregating within a population (Feldman et al. 1997). Consider two genes, A and B, each with two different forms or alleles, 1 and 2. Suppose that the fittest genomes are those carrying $A1$ and $B1$, and those carrying $A2$ and $B2$, and that the two alternative configurations (i.e., $A1$ with $B2$, and $A2$ with $B1$) are less fit. Clearly then natural selection will tend to increase the frequency of the favored two-gene DNAs and decrease the frequency of the less fit combinations. Importantly, note that when the $A1$-$B1$ and $A2$-$B2$ combinations become common, then recombination between them will tend to reduce their frequencies and to produce the other unfavorable types. Thus, natural selection and recombination would be acting against one another. Necessarily then, natural selection would also favor a mutation that acts to reduce recombination between the genes. The point is a general one. Whenever recombination is occurring and is producing an overabundance of recombinant genotypes that are less fit than the parental ones, then that recombination process will be discouraged by natural selection and mutations that reduce recombination will be favored by natural selection (Kondrashov 1988).

As mentioned above, there are two different modes of recombination to consider, and their differences are especially revealing in the context of the limits on recombination rate. Homologous recombination is when a portion of a DNA is swapped with, or replaced by, a portion of another that has a very similar DNA sequence. Recall that HOMOLOGOUS, when used in regard to DNA sequences, means that the two DNAs share a gene tree history and common ancestry (chapter 6). HOMOLOGOUS is a little bit of a misnomer in this context, as the actual criteria for this kind of recombination is sequence similarity, which for short regions of DNA can arise for reasons independent of common ancestry. In reality, two sequences are exceedingly unlikely to be very similar for reasons other than common ancestry unless they are very short. The second kind of recombination is incorporation, when a portion of DNA is taken up by another such that the final product is the sum of the lengths of the two participants.

Under homologous recombination, a DNA gives up what it was linked to, and so this mode of recombination should be sensitive to the kinds of interactions among polymorphisms that have been shown to lead to the reduction of recombination (Feldman et al. 1997). But under recombination by incorporation, a receiving DNA does not lose anything, but only gains. There could well be negative side effects caused by length, but the disruption of the interactions among portions of DNAs is less likely to suffer as a direct result of incorporation, as compared to homologous recombination. In general, we expect there to be greater limits to homologous recombination, than to incorporation, and this is exactly what is observed. Bacteria engage in homologous recombination at rates that are strongly dependent on sequence similarity (Majewski and Cohan 1998; Matic et al. 1995), but they are far freer to incorporate DNAs from a wide diversity of bacteria (Lawrence and Ochman 1998; Ochman and Bergthorsson 1998). Perhaps the best examples of highly promiscuous recombination via incorporation are the many cases where virus and transposable element DNAs have become incorporated into the genomes of extremely diverse host organisms. To list a few examples: double stranded bacteriophages have moved among a wide variety of bacterial hosts (Hendrix et al. 1999); members of the P-element family of transposable elements have repeatedly crossed species barriers within the genus Drosophila

(Clark and Kidwell 1997); retrotransposons have moved among sea urchins (Gonzalez and Lessios 1999), as well as among birds and mammals (Martin et al. 1999); and elements of the mariner transposon family have moved across entire phyla and kingdoms (Gueiros-Filho and Beverley 1997; Robertson and Lampe 1995). As these small DNA genomes do not greatly rely on collaboration with host cell genes, they are often viewed as parasites of the host cell. Promiscuous recombination via incorporation, that does not obey the boundaries that typify homologous recombination, is exactly what is expected for small self-contained sets of genes that do not interact strongly with the genes among which they become incorporated.

The Effect of Recombination on the Boundaries of Evolutionary Groups

Now consider again the patterns and processes that can arise within and among evolutionary groups. In the absence of recombination, evolutionary groups are expected to exhibit a fractal hierarchical structure and to have indistinct boundaries between groups. But suppose there is recombination. Would we then expect something other than fractals and indistinct groups? If recombination were entirely promiscuous, and occurred among all DNAs, we would not expect it to make these groups more distinct. In fact, recombination would remove whatever boundaries existed, and whatever fractal structure that existed as well. Promiscuous recombination among all DNAs would presumably leave just one large group. Furthermore, there being only one group, variation could not arise between groups.

Suppose, instead of promiscuous recombination among all DNAs, that recombination does not occur equally among all DNAs. This is what we expect if recombination evolved in a context of multiple evolutionary groups. Under these circumstances, individual DNAs would be much more similar to some DNAs, than to others, and recombination is not expected to be favored by natural selection between highly dissimilar DNAs. As noted above, this tendency for recombination to be restricted to within evolutionary groups would be especially true for homologous recombination. Here we can see how recombination might act as a lens on the fractal and indistinct structure of evolutionary groups. It will act to redistribute variation more evenly among DNAs that engage in recombination and thus to sharpen the divisions between those DNAs that cannot share in recombination. Recombination within an evolutionary group will also limit the rise of hierarchical structure within the group, thus dampening the tendency toward groups within groups. But between groups that cannot exchange DNAs, divergence can accrue without end. From this minimalist perspective, we see that the force of recombination holds the potential to turn a fractal landscape of great but uncountable diversity into one where the diversity lies primarily amongst a countable number of distinct groups.

But how strong will be the focusing effect of recombination? Certainly we can envision that at the limit of zero recombination, two evolutionary groups can diverge from one another. Thus for example whenever geographically separated evolutionary groups diverge to the point that they can no longer exchange genes, then that separation will, ipso facto, not become undone by gene exchange should the two groups come back into contact with one another. This is, of course, the basic model of speciation under allopatry (i.e., when populations are completely geographically separate). But among those DNAs that are not allopatric of one another and are capable of exchanging portions, expectations are less clear. By itself, the simple model does not obviously lead to predictions of just how low gene exchange must be to permit divergence between groups, or how uniform it must be to even out the levels of variation within groups. However, we can draw on some population genetic theory to understand this issue. A famous finding of classic population genetics, due to Wright, is that two populations cannot appreciably diverge if they exchange genomes at a rate much higher than an average of one per generation (Wright 1931). Though in the simplest version, the model ignores natural selection, it does provide us with the basic expectation that even a limited amount of exchange can make a large contribution to preventing divergence of populations. If we continue to set aside natural selection for the moment, then we can say that there should be just a fairly narrow range of low rates of recombination wherein indistinct boundaries and nested fractal structures could be expected to arise. When genomes occur as a recombining group, within which the rates are above this threshold, then a nested hierarchical fractal pattern of groups within groups should not arise. Furthermore if this group has a boundary, in terms of rates of recombination with other genomes that are below the threshold, then this boundary will also become manifest in patterns of divergence between groups of DNA sequences. Under these circumstances, an evolutionary group would be mostly distinct, even if there is a small amount of gene exchange with other such entities.

We see that recombination may, in theory, be expected to evolve and to also contribute to sharper patterns of diversity, that is, one in which evolutionary groups are clearly separated from one another and that contain less hierarchical structure. But as much as recombination may have such a clarifying tendency, several questions persist about the magnitude of this effect.

One major question concerns the stability of circumstances where low levels of gene exchange are occurring among partially distinct evolutionary groups. If two evolutionary groups are only partly distinct, and can exchange genes, is the amount of gene exchange expected to increase (and thus merge the groups) or to decrease (and leave two distinct groups)? One theory is that natural selection will tend to reduce the amount of gene exchange, by limiting the formation of hybrids, and so act to sharpen a boundary of limited gene flow. The process whereby a partial barrier to gene flow grows stronger via natural selection against hybrids was first described by Alfred Russell Wallace and is sometimes called the WALLACE EFFECT (Grant 1966). The same process was also called REINFORCEMENT by Dobzhansky (1940). If fertile hybrids form between partially divergent populations, but have reduced fitness relative to non-hybrids, then there will be an advantage to genomes that avoid forming hybrids, and thus additional barriers to hybridization might evolve. Indeed, there is good evidence from Drosophila (Noor 1995) and stickleback fish (Rundle and Schluter 1998) that this process does occur, though it may not be common. What is tricky about the model is that natural selection also acts among the hybrids, and the most reproductively successful hybrids may succeed in passing on genes that improve the fitness of hybrids. They will also be moving genes among the two populations, and thus removing the genetic differences that cause low hybrid fitness. For these and other reasons, the circumstances that permit evolution of reduced rates of hybrid formation are quite restrictive (Butlin 1989; Felsenstein 1981; Sanderson 1989).

We must also consider the effect of gene flow and recombination upon the boundaries of evolutionary groups when those processes occur at varying rates across geographic distances. Suppose that an evolutionary group includes organisms across a broad geographic range, such that gene exchange is effectively limited by distance. In circumstances like this, we can anticipate that a critical quantity will be something like the ratio of two distances. The first is the distance that DNAs move, either along with their organismal phenotypes, or as part of gametes (e.g., plant pollen in the air or invertebrate sperm in the water column). The second quantity is the distance between genomes that could re-

combine if the DNAs could be brought together. If the ratio of the
first quantity to the second is high, then genomes that could potentially
recombine will regularly have the opportunity to do so. But if the ratio
is low, then rates of gene flow will be low across a range of distances
that include potential recombination partners. This latter situation is
one of isolation by distance (Wright 1943), and it is one in which
DNAs and their descendants can become more divergent the farther
apart they are. In its simplest form, the isolation by distance model actually predicts that a pattern of uniformly dispersed organisms will be
replaced by a clumped distribution, as a byproduct of limited dispersal
and random successful replication (Felsenstein 1975). But if a density
dependent force is acting to distribute organisms more uniformly
throughout a geographic area (such as differential survival or migra-
tion), then it is not at all clear that distinct evolutionary groups will
emerge. Under these circumstances, each portion of the genome will
have a history of replication that may seem to reflect the presence of
one or more evolutionary groups, but there will be little correspondence among the histories of unlinked portions of the genome. Thus,
when organism are evenly dispersed, in the absence of selection against
recombination, one part of the genome may come to exhibit multiple
groups in some sort of geographical mosaic, but an unlinked portion of
the genome will come to have a different mosaic that only partially
overlaps the first.

Very much related to the idea of isolation by distance, is that of hierarchical population structure. Recall that in the absence of recombination, there is a basic expectation of fractal structure, of hierarchies
in the degree to which DNAs share in the processes that cause evolutionary groups (chapter 6). But it turns out that even with lots of recombination, hierarchies of evolutionary groups within evolutionary
groups may not be dispelled. The basic reason that hierarchies arise
within evolutionary groups is that competition, adaptations and genetic
drift tend to be shared most by DNAs that are near one another. Recombination, though it may spread genes around, is also something that
will occur at higher rates among DNAs that are near one another. Thus
recombination will not necessarily remove hierarchically nested evolutionary groups. As Dobzhansky recognized, obligately sexual organisms
often occur in populations that are hierarchically nested within one another, with hierarchical structure imposed by varying levels in the rates
of gene movement (Dobzhansky 1950, 1951). Finally it should be re-
emphasized that when hierarchically structured diversity arises by the

action of evolutionary forces on groups of related organisms and DNAs, those groups may well be neither countable, nor distinct.

The Distinction of Recombining Species

Now that the theory of evolutionary groups has been extended to include recombination, let us draw some more comparisons to species. We can only do this in a very rough way, for the evolutionary group model is an argument, and it is not quantitative. It does not give us any numbers with which to compare with data on species, nor was that the reason for devising it. Rather, the purpose was a qualitative comparison, to see whether a simple model, based on replicators and competition, gives rise to the kinds of phenomena we find with real species. What the model says is that evolutionary groups will arise, but that they may not have distinct boundaries. It also says that recombination can be expected to evolve and that it will act to promote sharper boundaries between groups and to reduce hierarchies within groups. But it was also noted that the model did not necessarily predict that all boundaries would be sharp or that all hierarchical patterns would be removed. We also do not know what species are (the species problem), so the comparisons that we can draw focus mostly on how the evolutionary group model corresponds to some of our uncertainties about species.

Certainly the evolutionary group model is consistent with the rise of groups that are distinct from one another, even if each group itself might contain multiple groups. Two evolutionary groups that do become different, so as not to exchange genes or to compete directly, can both persist and continue to diverge. This can occur even if each group includes multiple indistinct groups. Thus the model allows for that most pressing aspect of biological diversity that Dobzhansky described (chapter 6) and that appears to us as a vast number of different natural kinds of organisms. But for our purpose, the more interesting comparison between the evolutionary group model, and species, concerns those cases where biologists have studied one species well, or have studied what they think may be two or more closely related species, and are uncertain about boundaries and species counts. The simple evolutionary group model quite liberally allows for partial boundaries and fractal hierarchies within evolutionary groups, even ones where recombination can occur. As most biologists who have closely studied species know, these are also common features of species, and when they occur they contribute to our uncertainty about species.

In the first place, hierarchies of relatedness have long been recognized as a fairly ubiquitous fact of life, both among distinct kinds of organisms and within well-studied species. Dobzhansky's view of species is especially telling in this regard, for he was a strong proponent of the idea that species are groups of organisms bounded by a capacity to reproduce (see the Biological Species Concept, below). But he also recognized that within species, that appear to be distinct from others by the criteria of reproductive behavior, there are hierarchies of levels of competition and genetic drift, and that hierarchical patterns of reproduction and gene exchange contribute to these hierarchies (Dobzhansky 1950, 1951). Perhaps the basic idea is most easily pictured by considering humans which, despite their mobility are still geographically constrained in their habits and still are much more inclined to reproduce with nearby individuals than with distant ones. Thus on average, individuals that are near one another compete more directly for resources and for mates, and this contributes to a roughly nested pattern of genetic relationships. To note just a few other recently described examples, hierarchical patterns of groups within species have been reported for Artic Char (Brunner et al. 1998), fire ants (Ross et al. 1999), and a fungal pathogen of trees (Gosselin et al. 1999).

Second, biologists have long been interested in just how distinct are species, and there is a long tradition of debate on, and investigation of the relative frequency and importance of recombination and gene exchange between individuals of different species. Fertile hybrid offspring are known to occur for a great many groups of closely related species, and not surprisingly these hybrids have played a central role in many species problem debates. Despite the many reports of hybrids, biologists disagree on what proportion of species might actually engage in hybridization and just how to judge the importance of such hybridization. Thus Mayr (1963) emphasized for animals, particularly birds, that hybridization is a rare event, while Grant and Grant concluded that it is a common phenomenon among bird species (1997). In the case of plants, where many have noted that hybrids are quite common (Anderson 1949; Arnold 1997; Grant 1981), some authors have still concluded that hybridization is rare and unimportant (Mayr 1992; Wagner 1970). This is surprising in part, because plants clearly harbor a great many cases of hybridization between recognized distinct forms (Arnold 1997; Harrison 1990; Rieseberg and Wendel 1993), not to mention the cases where distinct new species arose via polyploidy (i.e., by the perpetuation of multiple chromosome sets in descendants of a species hybrid). These latter cases are clear examples of speciation by hybridization, and they occurred at various times in the phylogenetic history of most flowering plants (Arnold 1994; Grant 1981; Masterson 1994).

Historically most of the evidence of hybridization between species has been through the observation and counting of hybrids that were identified on the basis of morphology (the shape of an organism) or cytology (chromosome number and shape). The most direct way, observation of organisms engaged in reproduction in nature, is very difficult and nearly impossible as a means for assessing low levels of hybridization for most organisms. The striking exception has been Darwin's finches on the island of Daphne Major in the Galapagos. In this case, long-term and intensive observation by Peter and Rosemary Grant and colleagues has shown that some species hybridize with others at a rate on the order of 1%. Importantly, the hybrid progeny are fertile and reproduce apparently with success similar to nonhybrids (Grant and Grant, 1997, 1998).

Today, most of the evidence for gene flow between species comes from the analysis of the histories of parts of their DNA genomes. In their most complete form, these studies present gene tree estimates from numerous organisms in each of multiple closely related species *and* for multiple unlinked portions of the genome. Still useful, but less thorough, are multilocus studies of allelic diversity, again within and between closely related taxa. Allelic variation, such as at blood group genes, or protein electrophoretic variation, or the more recently discovered short tandem repeat loci (also called microsatellites), does not permit the same level of resolution as DNA sequence based studies. But one can often still assess whether all loci are consistent with a common model of divergence. What these multilocus studies permit is a view of variation across the genome in how two species have diverged. The critical question is not how much divergence has accrued, but whether it has done so at similar rates and in similar ways at different parts of the genome. In several cases, there is evidence of gene flow for some genes, but not for others, or that rates of gene flow vary among portions of the genome. Findings like these suggest that species barriers, and thus species as well, are truly distinct at some genes and truly indistinct at others. Some examples from plants include sunflowers (Rieseberg et al. 1996a, 1996b), endemic and introduced species of *CARPOBROTUS* (Gallacher et al. 1997); red and white campion (Goulson and Jerrim 1997); Australian cotton species (Wendel et al. 1991), oaks (Samuel et al. 1995) and buckeyes (DePamphilis and Wyatt 1989). Some other examples of taxa where different genes appear to have had different levels of gene flow between closely related species include: enteric bacteria (Lawrence and Ochman 1998); Daphnia (Taylor and Hebert 1993); Drosophila (Hilton et al. 1994; Powell 1983; Wang et al. 1997); field crickets (Harrison and Bogdanowicz 1997); grasshoppers (Shaw et al. 1979); mosquitoes (Kamau et al. 1998; Lanzaro et al. 1998); spiders (Piel and Nutt 2000); grass shrimp (Garcia and Davis 1994); swallowtail butterflies (Hagen and Scriber 1989, 1995); mussels (Heath et al. 1995); land snails (Clarke et al. 1996; Porter and Ribi 1994); mice (Ferris et al. 1983; Vanlerberghe et al. 1988); cyprinid fish (Dowling and DeMarais 1993); and trout (Giuffra et al. 1996). Partial and uneven species boundaries are common, perhaps even the rule for closely related species, given the frequency with which they are revealed by genetic studies in recent years.

The simple model of evolutionary groups was devised to provide a framework with which to help think about how patterns of diversity may arise among replicators. The framework provides a helpful context for thinking about one of the most difficult things that biologists have discovered about species—and that is their often fuzzy nature. Though we might well expect evolution to give rise to real groups of replicators that are enjoined by shared forces of evolution, we also expect groups to have indistinct boundaries with respect to closely related groups, and to contain other smaller groups nested within them. Turning to actual reports on species, we find, of course, that these same kinds of observations have been made many times. There should be no surprise in this. Biologists have long known, from their models of speciation, and from studies of divergence, that barriers to gene exchange between evolutionary groups must often evolve fairly slowly, both when incipient species are allopatric and when they are sympatric (Coyne and Orr 1989; Grant and Grant 1997). But still, many species debates overlook the simple fact that real species may exist in nature, and yet may still be partly indistinct and uncountable. Just as with clouds (see chapter 2) or indeed with many other entities that are dynamic and made of intermovable components, the fact of existence is note undone by the presence of indistinct boundaries.

The Biological Species Concept

The reduced view of recombination within evolutionary groups also leads us, through the back door, to the Biological Species Concept (BSC). The BSC was intended as a definition of the word SPECIES and also as a description of the criteria to be used in recognizing species.

Though the exact words that serve as the definition of the BSC have changed over the years (Mayr 1942, 1982, 1996), there is a clear core that holds that a species is a group of organisms that reproduce with one another and that are not capable of reproduction with the members of other such groups. One recent version is that "a species is an interbreeding community of populations that is reproductively isolated from other such communities" (Mayr 1992, 222). Another is that "Species are groups of interbreeding natural populations that are reproductively isolated from other such groups" (Mayr 1996, 264).

The main point, that organisms occur naturally in groups that are delimited by interbreeding, and lack thereof, is a great idea from several vantage points, and despite being the subject of much criticism, it is one of the most extensively utilized notions in all of the history of biology. Today, the names most associated with the BSC are Dobzhansky and Mayr. It was Dobzhansky who explained speciation in terms of genetics and who popularized the use of genetics for the understanding of species diversity (Dobzhansky 1937). It has been Mayr who strongly championed the role for these ideas as an explicitly named concept. Both scientists also made careers of studying species, Dobzhansky more from the standpoint of genetics and Mayr more from the biogeographic side. The ideas behind the BSC are, of course, not 20th-century innovations. As early as 1749, Buffon laid out a description of what we today call the BSC, and proposed it as the defining criterion of species (Mayr 1982, 334). Mayr (1982, 263) also describes how these ideas were commonplace among biologists in the 18th and 19th centuries. For what it is worth, the seeds of the idea also lie plainly in the myth of Noah's ark.

The BSC is somewhat atypical among species concepts for its strong theoretical component. Not only is the concept directed at the question of "What are species?"; it is also partly an explanation of the causes of species. The theoretical focus leads investigators fairly directly to research programs on the determinants of patterns and capacities for reproduction. In principle, and to some degree in practice (as we have seen above), the ideas behind the BSC can be used to gather evidence that species are not distinct or do not exist. To the extent that one does not find groups of organisms that are delimited on the basis of reproduction, then quite simply that criterion does not support a model in which organisms fall into distinct reproductive groups. In short, one could embrace the BSC without making a strong species presumption.

However, the BSC is also very problematic, and many authors have commented on the practical difficulty of the BSC as a guide in identifying or classifying organisms (taxonomy) or as an aid in figuring out the evolutionary relationships among organisms (systematics). As we shall see in chapter 12, the ultimate reason for this is confusion over the purpose of a species concept. In this confusion, the BSC shares all the troubles experienced by most other species concepts.

Evolutionary Groups and Species

Because the theory of evolutionary groups follows from the ideas of molecular replicators under competition, we can use it to return to a question that was raised in discussion of the theory of life. Recall that a basic version of the theory of life described in chapter 3, which very roughly explains how life evolved, did not include species. The pressing question was whether the theory of life was deficient for not including species: "Is evolution necessarily something that happens only to species?" We don't yet know what to make of species, and cannot answer the question in those terms, but I think we can say that evolutionary processes will tend to be shared by groups of related replicators. Evolutionary groups are where replication, competition, natural selection, mutation, and recombination happen—all the usual suspects of evolution. These are the processes that underlie those most visible outcomes of evolution, the new adaptations and the new kinds of organisms. Indeed there are scarcely any theories of adaptation, or of the origins of new kinds of organisms, that do not work by precisely these mechanisms. The noteworthy exception is that sometimes new species can be formed suddenly by a single hybridization event between just two organisms, one from each of two other species (Arnold 1997; Stebbins 1950).

But even if evolution really is a process that is caused by multiple competing replicators (or organisms), why has all of this discussion been written using EVOLUTIONARY GROUP and not SPECIES? The reason is a simple wish to avoid some of the confusion that can arise with SPECIES. The purpose of this book is not to define SPECIES, but rather to explain why we have such difficulty with the word. EVOLUTIONARY GROUP is a term without the baggage of generations of confusing debate. The ideas behind the notion of an evolutionary group are in some ways entirely conventional, but the concept and the argument behind it are a reinvention of thinking about biological diversity. The ideal is of an understanding that is not jinxed from the outset by the species problem. We can hold on to and use this understanding of evolutionary groups as a way to refer to patterns of biological diversity that does not require invoking SPECIES. But, of course, there is a very close connection, one that is strongest when we envision species as real groups of related organisms that share in an evolutionary process (as biologists commonly do). In the remainder of the book, I rely on EVOLUTIONARY GROUP, but I also sometimes use REAL SPECIES as a synonym. A summary of the connection between the ideas behind these two words and recommendations regarding their use are provided in the penultimate chapter.

EIGHT

THE CAUSE OF THE SPECIES PROBLEM

The pieces are in place, and we can now join the results of the back-door approach to species, as biological entities, with the results of our backdoor approach to categories in language. By starting with the theory of life, we have built a model of how evolution gives rise to groups of DNAs and organisms that share in competition, and thus also share in genetic drift and adaptation. This idea of evolutionary groups corresponds closely to a major theme that is common to various definitions of SPECIES, which is that a species is an entity within which the forces of evolution occur. We have also explored the existence of categories, particularly the difficulties that arise if categories are taken as literally real, outside of ourselves. For pragmatic scientists, natural kinds and other categories are different from entities, and are not real in the way that entities are, though they do exist in the minds of those who devise and use them. The psychological investigation of categories have revealed that the mind wields a natural kind by drawing on some kind of internal representation of a typical, central member. Finally, anthropologists and psychologists have shown that people do this for kinds of organisms as well. Whatever else species are, they are quintessential natural kinds and they exist as categories in our minds, complete with prototypes.

If this synopsis is a roughly accurate model of what really happens in the world, and of what really happens in our minds, then we've got an odd situation. What is weird is that many of our species, by which I mean our mentally constructed natural kinds of organisms, actually do correspond to real entities in nature. Consider, for example, how the person category is one of our most familiar and frequently used natural kinds. But, in fact, all the currently living things that we would include in that category actually constitute an evolutionary unit, a real evolutionary group that we know as the human species. What is clearly a conventional category in the mind also has the unusual property of somehow corresponding to an entity in the real world. This is not the case for most other natural kinds. All snowflakes do not make one thing, and neither do all kidney stones, or lighting bolts, or all of pretty much any category. In contrast, the person category is doing double duty, with chores that span a major ontological chasm. As a category, we use it to contain all living persons and also sometimes, at our whim, it includes a great many other things, like George Washington and Tarzan, that either no longer live or never lived. But PERSONS, or PEOPLE, is also a common name for a real evolutionary group, a symbol that we use to refer to that entity.

Note that the semantic duality of a name for a species is not a matter of homonyms. PEOPLE as a name for a category and as a name for an entity is not at all like, say, BANK for both river shoreline and financial institution. Nor is it like JAZZ, which is both a category (of music) and a name for an entity, the Utah Jazz, a basketball team (formerly of New Orleans, where jazz, the music, was first played). In the case of PEOPLE, as well as the names of other species, the two different meanings are somehow very closely connected to each other in our minds, even though they convey entirely different modes of existence. To find parallel examples that are not evolutionary groups, we need to think of entities that consist of multiple similar components. For example, as of this writing, Orlando Hernandez, the baseball pitcher, is a Yankee, and he is so in very much the same way that he is a person. We place him in the category of baseball players who play for the Yankees, and that baseball team is also an entity—though perhaps not very distinct, it is made of interacting components and is somewhat constrained in space and time.

The dual nature of species has caused some consternation when considered in light of the nominalist/realist difference (chapter 4). Recall that Plato and Aristotle held that kinds of things, categories found in the real world, held a real essence. But philosophical nominalists countered, persuasively, that categories are not real because they cannot do things and they cannot be acted upon. This *either/or* debate also extended into biology, for just as Linnaeus embraced the essentialist (i.e., realist) view of species, so too did nominalists come to question this view. The reasoning was simply that since categories are not real, and since species were quintessential categories, it followed that species were not real. This view was expressed in various biological contexts in the 18th and 19th centuries (Mayr 1982, 264). Indeed, the two key points, that kinds of organisms are categories and that categories do not exist, are so salient that some 20th-century authors have still been lead to conclude that species do not literally exist (Burma 1954; Gregg 1950). In doing so they overlooked the clear counter argument that was to follow, that species as envisioned by many evolutionary biologists, are entities composed of multiple organisms that share in evolutionary processes (Ghiselin 1966, 1974; Hull 1978). Michael Ghiselin (1997) is most direct on this point, insisting that species are real things and are not categories. Unfortunately Ghiselin's and others' embrace of the nominalism/realism clash has focused on the traditional dichotomy. The assumption that it is an *either/or* problem is both unnecessary and misleading. Ghiselin is certainly correct that there are evolutionary entities that deserve being identified as species, but regardless of the reality of species there is no getting around their existence as categories. We have millions of named species, and whether or not any or all of them match up well with real evolving entities out there in nature, they are all categories, each and every one has been devised by a person or persons as a way to help organize our understanding about biological diversity.

All that is needed, for a person or people to devise a category, is recurrence. People are very good at noticing when similar things happen more than once, and they don't balk at devising categories to handle such additional recurrence as may happen. In the case of evolutionary groups, it just so happens that the forces that make such a group also cause the components of that group to be similar of one another. We notice and are impressed by recurrent, similar organisms, and these impressions cause our categories of organisms, complete with prototypes. Indeed for most times, and for most people, there is little reason to move beyond recurrence, to ponder or question the causes or nature of these recurrent patterns. Having a word, say CHICKEN, with a corresponding idea of a kind of animal or a kind of food, will pretty much serve most people's needs any time they must deal with yet another instance of that category. That is what categories are for—they are inspired by recurrent things, and in turn they allow us to refer to whatever additional recurrence turns up. There is more to say on recurrence and categories in chapter 9, but for now it is sufficient to emphasize that the construction of categories is a basic human imperative. Some people today might pass large periods of time without even seeing, or otherwise perceiving another organism that is not a person. But consider the cook, or doctor, or farmer, or hunter, or biologist whose livelihood hinges in part on knowledge of many kinds of organisms. These people will have an inkling of the lives our ancestors lead, and of the lives of present day hunter-gatherer people, for whom a vast part of their knowledge about the world lies in their categories, their kinds of organisms. Life for one of our distant ancestors would be fairly meaningless, semantically and thus literally, if it did not include a strong ability and willingness to devise and wield categories of organisms. That life would also probably have been a short one, as organismal categories must have been one of our most basic survival tools. Finally, be sure to appreciate that all of this categorization can proceed without the slightest inkling of the causes of recurrent kinds of organisms.

But the development and usage of categories is just our most facile and immediate response to recurrence among organisms. The patterns of biological diversity have also evoked another, very different, human behavior that is a scientific response—a drive to understand the causes of that diversity. Our curiosity has lead to a rich understanding of evolutionary processes and of the origins of life, one that includes the reasons why some organisms are so recurrent of one another. We have learned that the patterns of similarity among some genomes and organisms are caused by evolutionary processes that go on amidst closely related, closely competing genomes and organisms. We know too that the dissimilarities that inspire our many different categories of genome and organism are caused by many separate diverging evolutionary processes.

In brief, modern biologists suffer two imperatives. The first is the ancient one of all people and that is to devise categories and invent just as many kinds of organisms as we want or need to give voice to our thoughts about that diversity. The second is to understand the causes of that diversity. Indeed, our pursuit of that second imperative has been so successful that it has given us a species problem. Back when Linnaeus promulgated his incorrect theory of species—that they were real kinds with real world essences created by God—the match between real species (supposed) and mental categories was of a one-to-one correspondence. Linnaeus' was not a theory of categories, but today we can appreciate the close correspondence between the way people hold kinds in their minds with the way that Linnaeus supposed they exist in the world.

But since the time of Linnaeus, our improved understanding of the causes of diversity has given us a rift between our theory and our categories. We now know something of what makes a real evolving entity in nature. But the more we search for those entities and study them, the more we find a poor correspondence with our categories. We try to fill the breach by debating our definitions of species and of how to identify species, but to little avail. As if a bull bearing horns, we have rushed with our twin imperatives—to categorize and to understand—and impaled ourselves in a place that we don't easily see. We feel the pain of the species problem, but unable to see ourselves as the cause, we press and we fret and we argue, with little alleviation.

Of course, there is an ultimate and profound connection between evolutionary groups and our categories. Our theory of evolutionary groups describes the entities, the evolutionary role players, wherein similarity and divergence are created. Within groups, the processes of evolution cause similarity, whereas between groups divergence accrues over time. Thus we are perfectly correct to say that evolutionary groups are the cause of our categories. But they are an ultimate *cause* of the categories; quite frequently they do not match the categories. Put another way, evolutionary groups cause the similarities, but the organisms that we include in a category very often do not constitute an evolutionary group. There are three related reasons why evolutionary groups are often not the proximal subjects of our categories:

1. Evolutionary groups are entities of the moment, existing because of what goes on among their components. However, much of the similarity that is generated among the members of an evolutionary group may last long after one evolutionary group has diverged into many.
2. Our perceptions are not a perfect mirror of nature. We observers can only employ perception and judgment and those are mental processes that lie within us. At base, these abilities are adaptations that served our ancestors' goal of procreation. As capable as we are, we must also appreciate that we are not omnipotent. There is no reason why our senses should be as capable of the same subtleties as can arise in nature.
3. Evolutionary groups need not be distinct, and can be nested within one another, whereas categories are typically wielded with an all-or-none-ness that comes with having distinct words.

So finally we can sum up, in a few words, the cause of the species problem. Here it is in two sentences: (1) Biologists have been looking for a theory that explains the species they recognize and that can be used to define SPECIES, and we have assumed that the theory will be about things that match up with our categories, our kinds, of organisms. (2) However, the entities that are in the best theory, which does indeed explain our kinds of organisms, need not and often do not match up with those kinds.

Please note that by "best theory," I don't just mean the theory of evolutionary groups outlined in chapters 6 and 7. As I've said, the key idea of groups of organisms that share in evolutionary processes, is one that many biologists have seen as a basic component of a species concept, and many species concepts are strongly consistent with this idea. Biologists who are persuaded of this view also freely recognize that real evolving entities need not be distinct and need not be easily recognized. As discussed in chapter 2, biologists are actually quite familiar with how poor and inconsistent are their measurements of species. What has been missing is a failure to appreciate that an evolutionary theory can explain kinds of organisms, and yet at the same time, the entities that are in that theory need not correspond to those kinds.

The Categorical Necessity

But why have biologists missed this? Why have we not seen that our tendency to focus and rely on our categories has distracted us from the real action? I think the reason is our very strong reliance on our categories, a reliance so strong that we tend not to notice that our categories lie largely within us. As much as our categories are caused by recurrence in the real world, they are also caused by our own interpretative apparatus. That apparatus must have evolved as part of the more general capacity for language. Within the capacity for language lies the ability to recognize recurrent aspects of nature, to embody within the mind some sort of representation of typical aspects of that recurrence, and to devise words so as to refer to instances of that recurrence. The typological method behind our capacity for recognition and reference almost certainly arose as a byproduct of the typological nature of all evolutionary adaptations. These points are discussed more extensively in the next chapter, but whatever its origins, we use our capacity for categories to organize our perceptions and language about nature. It is a behavior we have, and one that must surely be rooted very deeply. Finally, it is a behavior that offers no particular guarantees of revealing the real entities that cause our categories. Our categories of organisms are good tools in many ways, but they come up short in the face of modern demands that they match our understanding of evolution. Even as our categories fail us we still cling to them, which is another way of describing the cause of the species problem.

NINE

THE ORIGIN OF NATURAL KINDS

Suppose that we would like to overcome our categorical bias toward understanding organismal diversity. How should we proceed? We cannot simply give up our category-filled language about organisms. The final chapter lays out some suggestions for how to deal with the species problem, so as to better promote our understanding of biological diversity, but they are not radical. Even if it were desirable, a self-imposed right-think code of speech and thought about species is probably impossible. It is hard to even imagine what such a code would look like. For this reason, and for those outlined in this chapter, I suspect that our mental functions of conceiving and interpreting the world are grounded in very basic mental tasks. But why might this be the case? Is there any particular reason, either psychological or biological, to suppose that our reliance on categories is old and deep? I think there is. The purpose of this chapter is to show how the representational mental method, that gives rise to prototype effects and that seems to be behind the categories of out thoughts and language, could have arisen. It describes the hypothesis that typological categorical thinking arose as a necessary byproduct of the way adaptation happens in a world of limited recurrence. If roughly true, then our reliance on categories, and our prototype effects may be both ancient and fundamental to the way we organize our thoughts.

To get the argument started, adopt for a moment an imagined omnipotence, as if you can see the entire universe at all scales and times. Imagine that this view includes: the energies of the universe; and also all the atoms and their parts, bridging as they do in the case of parts of atoms, the existential extremes of probability and entity. Imagine as well that you can see all the molecules zooming and wiggling; with the slower and bigger molecules often clumping in large aggregates, themselves exhibiting a nearly infinite array of properties. The reason for adopting such hypothetical mastery is to help introduce, in a very general way, the notion of a circumstance in the world. A circumstance in this context is any localized matter, energy, or interaction (e.g., process) that happens. What's more, as an omniscient observer one has no tendency to circumscribe a circumstance by one's own perception. A circumstance might be seemingly self-contained such as a photon winging its way, or a grain of salt dissolving in a raindrop, or a child sitting in a chair eating a sandwich. But if we are omniscient, then there is no such containment, and we are as likely to perceive a slice of any of those examples as a circumstance as well. While it may seem trivial to introduce the notion of a circumstance and to say that circumstances occur, it is less easy to imagine a view of reality that is not bounded by perceptions of boundaries.

We also need the idea of circumstance in order to introduce the next idea, which is recurrence. Recurrence is when a circumstance is reiterated as a matter of degree, when something like it happens at another time or place. Recurrence is the phenomena that necessarily underlies all natural kinds, whatever their ontological status and whatever the psychology of our references (Landesman 1971).

To appreciate the causes of recurrence, we should first recognize that all circumstances must necessarily contain smaller circumstances (excepting, perhaps, those in the quantum limit of smallness). Thus in a somewhat trivial and only proximal sense, the primary cause of recurrence is recurrence among component circumstances. Two circumstances can only be reiterations of one another to the extent that they resemble one another, and similarity necessarily requires some commonality among the two sets of components. This nesting aspect also reveals the second proximal determinant of recurrence, and that is size. The circumstances over a short distance and time are more likely to be similar than those over longer scales, and this is for the simple reason that recurrence for a large circumstance implies recurrence for the many smaller ones within it. Thus, if we recognize ideas of circumstance and recurrence as a way to approach a discussion of patterns in the world, then we would also recognize that a basic feature of recurrence is disparity creep—two circumstances cannot be completely identical, and there is more and more disparity on average for larger and larger circumstances. A carpenter may build two apparently identical houses, similar to every detail within his grasp, and yet those two houses could not possibly be recurrent of one another to the same degree that, say, two molecules of water, or even two glasses of water, could be.

One flaw in this discussion of circumstances is the imagined omnipotent view under which circumstances are to be observed identified and distinguished one from another. In practice, there is not a true knowledge of circumstances, but rather a human interpreter who describes apparent circumstances and apparent recurrence. Humans recognize and give names to patterns in the world that seem to have aspects in common, and we are continuously observing, finding and giving words to, what seem to us as, recurrent circumstances. And unlike the omnipotent observer, we tend to notice circumstances as a function of boundaries that we perceive. Furthermore, we are tuned to note recurrence on the scale over which our perceptions operate. To us, two houses may indeed seem identical, while an omnipotent observer across all scales, or another observer at a finer scale (say a termite), might not be so impressed. Recurrent circumstances that we recognize, our categories of seemingly oft-repeated aspects of nature, and that we give names to, are the natural kinds discussed in chapter 4. These categories are within us, as some kind of distillation of recurrent patterns in nature. But they way that they exist within us is not simple and neither is it logical in a set-based theoretical way. A major point of chapters 4 and 5 was that our minds and bodies are not neutral observers and descriptors of recurrence, but that the way that we are built, and the ways that we perceive, interpret and ultimately refer to recurrent aspects of nature actually embody and represent those recurrent aspects of nature.

The next step is to realize that our categories are a reaction, a consequence, an effect, of recurrence in nature. If one stands on a hill and sees in the distance a very high peak, then any thoughts or words that are made about that mountain are an effect of that mountain: they are partly caused by that aspect of the landscape. Also, the use of a categorical word like MOUNTAIN and one's thoughts of the peak as an instance of a category are also an effect of that peak, as well as an effect of similar high peaks, other references to mountains, and the use of MOUNTAIN that has gone before. Recurrence is the cause, and a category is the effect. Though the cause and effect are not simply connected, still we must suppose that among the causes for thoughts and language regarding recurrence, indeed the major cause, must be that recurrence. Of course, humans engage in all manner of thought and language without explicitly directing them at a very instance before them. Idle thoughts, fiction, past and future references, are all probably a larger part of the human repertoire than explicit pairings of words with nature. But the genius and generality of human reference does not take away the fact that the ultimate cause of categories and natural kinds, as they exist in thoughts and language, is recurrence in nature. When Joyce Kilmer wrote "I think that I shall never see a poem lovely as a tree," his reason for writing the poem and his use of TREE must have had a plethora of causes. But there is no escaping that trees, or that recurrence in nature which we have come to refer to with TREE, was an ultimate cause. How could one suppose otherwise and still hold that words about nature have meaning?

To grease the wheels of the argument to come, let me reinforce a distinction that has been made and implied at various places in previous chapters. That distinction lies between the words used for what exists in nature, outside of ourselves, and those for what exists in thoughts and language. I will persist with RECURRENCE as it was described above, as meaning similarity, in a very general way, among circumstances in nature, and I will use CATEGORY and NATURAL KIND when I refer to the way that recurrence is represented in our thoughts and language. The distinction is partly artificial, for we are not omnipotent and we do not know reality apart from ourselves. Our language and thoughts are as much a part of the world as anything else. But it deserves repeating that we must make distinctions like these if we are to discuss language: we must assume some objectivity and capacity of language in order to explore the limits of that objectivity.

Recurrence, Thought, and Language

Now consider the processes of perception and thought that lie in between recurrence in the world and language that we put back into the world. We perceive and think about recurrence, and those processes lead in turn to language. I would ask a reader to imagine this very rough characterization as a chain of three steps: recurrence in the world; followed by processes in the mind and body; followed in turn by language that is put back into the world by the mind and body. We can even diagram it like so: *real world* $\Rightarrow$ *mind and body* $\Rightarrow$ *language*. But what kind of mapping actually occurs, both into and out of the mind and body? Is our internal representation a copy of the world, and does it give rise to an accurate representation of the world in language? These questions were partly explored in the description of the discovery of prototype effects, whereby categories in the mind seem to have a center with fuzzy boundaries, and instances can be either good or poor examples (chapter 4). The discovery of prototype effects played a key part in the overturning of the classical view of categories and the rejection of a simple correspondence between language and reality. Under that partly dismissed, but orderly view, the categories in language are based on mentally held rules for membership. But even if the classical view has been rejected as a model of the mind, it still seems to be a good description of language and the world. According to that view, the real world is made up of instances of distinct kinds (universals) and, befitting the idea that language corresponds to the world, so too does language appear to have distinct categories filled on the basis of distinct criteria of membership. As noted near the beginning of chapter 4, language is great for describing the world and it does, at least in a commonsensical way, seem to fit the world. But if, in fact, the classical model is a poor one for describing the mind that lies between the world and language, then we are left with a paradox. According to the revisionist story, under which the classical, rule-based models have been rejected, our underlying mental processes do something different that is both out of sync with the world and also out of sync with the seemingly straightforward way that grammar juxtaposes distinct entities and categories (Lakoff 1987). This tale of philosophers' and psychologists' changing view of categories assumes that language *does* fit the world, whereas it is our underlying thoughts, with their prototypes, that do not. It is as if we suppose an apparent relationship of strong commonalities (*real world* $\Rightarrow$ *language*) that belies a hidden complex relationship with many fewer commonalities (e.g., something like: *real world* $\approx>$ $\boxed{mind\ and\ body}$ $\approx>$ *language*).

It should become clear as the argument develops, why I think this story of an inaccurate intermediary mind and body, is wrong. It is wrong for overplaying the apparent similarity between language and recurrence, and it is wrong for failing to appreciate the greater similarity between mental prototypes and the recurrence that causes them.

Take note of a critical implication that is a consequence of two key aspects of recurrence that were described in the previous section. The first was that recurrence in the world is not absolute, that patterns of similarity also include a great deal of dissimilarity. The second point was that recurrence in the world motivates our use of categorical language. We devise and deploy words as a consequence of recurrence. Together these two points mean that our words do not, cannot, map onto the real world in any absolute way. A categorical word may be a very useful indicator for some highly recurrent patterns, but even so, those patterns are not completely recurrent. In short, the classical view in which the subjects and predicates of sentences match up with particular entities and universals, respectively, overlooks the fundamentally vague nature of recurrence in the real world.

Is there some reason to suppose that a category that is built around a central exemplar, a prototype, is inherently inaccurate? The answer is certainly "yes" in so far as real world recurrence does not literally have prototypes, as there are no truly better or worse instances of anything. Under an adopted omniscient view of recurrence, circumstances do not have boundaries, real world recurrence is vague and partial, and categories have no representation outside of the mind, either as essences or rules. It is when one is not omniscient, when an observer has a point of view, that recurrence seems distinct and bounded. Now suppose that, knowing recurrence to be truly vague, one had to design a method of having a point of view and of recognizing recurrence. In this context, prototypes make a great deal of sense. Would not a method that admitted fuzzy boundaries, with good and bad examples, be a better reflection of vague reality, than a method that lumped all interpreted instances of recurrence under identical criteria? In short, when an observer is not omniscient and does recognize recurrence from some point of view, then a method that allows for disparity is more accurate than one that does not. If so, then the odd one out among reality, thoughts and language, is language. Our grammar of discrete words—subjects and predicates, digitally encoded—is farther removed and more dissimilar from the continuous world than are our perceptions and thoughts about that world.

Where Categories Occur

It is sometimes overlooked, in the philosophical and psychological discussion of categories, that human interpretation of recurrence and our language of natural kinds are not the only realms for the recognition of recurrence and they are not the only places where categories can occur. Certainly human minds and human language together form a sort of category engine; but the causes and effects that are recurrence and categories, respectively, are not found only in human minds. Consider any organism that is capable of learning, where the organism holds the potential to behave in a certain way in response to future circumstances that resemble others that have been previously recurrent. Learning is a categorization process, and thus an interpretation of recurrent circumstances, and it is practiced by organisms with minds so simple that we don't typically refer to them as minds.

Consider too the actions of an immune system that learns to recognize foreign antigens. At the front lines of an immune response are antibody molecules that hold the potential to recognize antigens. And as with mental learning, the recurrence that gives rise to an immune response is not perfect. Certainly a perfectly uniform and identical set of antigens can give rise to a uniform and identical set of antibodies, but in the real world many antigens are somewhat heterogeneous. Thus, for example, the different particles that are agents of a disease (be they cells, viruses, or proteins) will be somewhat variable for various reasons, and the immediate circumstances in which they are presented to antibodies will vary (e.g., in local pH, temperature, and the concentrations of various moieties). The immune system will generate a response that is a distilled effect of the commonality among the circumstances when antigens were bound. But as with a learned response, an immune response has been caused by some circumstances more than others. It might seem strange to say that an immune system contains categories, or that it refers to recurrent antigens, but think of the parallels with the way a mind has categories. In both cases, when there is recurrence in nature, one effect of this recurrence is that a mechanism arises for reacting to additional recurrence. In the human mind that method is a category, and in the immune system it is cells bearing genes that make specific antibodies. In both cases recurrence has caused the capacity to recognize further recurrence. Finally, take note of another important parallel between the natural kinds recognized by people and those recognized by a simple learning organism or an immune system. In all these cases, there is no need—nor perhaps any possibility, strictly—of exact recurrence among the instances of the category. An organism or an immune system may learn, and a person may think of a category, and these processes are facilitated by more- (rather than less-) recurrent circumstances, but they are not dependent on strictly recurrent circumstances.

This cause and effect view, of recurrence and categories, and the realization that in this way human categories resemble other systems of recurrence and recognition of recurrence, is the thread that we will pursue. It turns out to be a useful idea. Indeed like a root that we pull only to find it part of a tree so large we somehow didn't notice it before, these parallels lead to some larger insights. Let us ask, where else in the biological world does it happen that some recurrence causes the recognition of recurrence? Consider Darwin's essential discovery, that natural selection gives rise to adaptation and diversity. We know that at base what is going on is the differential persistence of DNAs. There is a process of an increase in the number of some DNAs by virtue of the way they differ from, and are better at replicating than other DNAs. When this process of replacement, of less fit by more fit DNAs, occurs again and again, there is a steady emergence of a newly crafted, finely improved, evolutionary product. This is the process of adaptation. We also use the word ADAPTATION to refer to individual salient aspects of the genotype or phenotype that we recognize as a component result of this process (e.g., as in "an adaptation"). These ideas are straight out of Darwin (though cast here in terms of DNAs), and Darwin well realized that adaptation requires regularity of circumstances. He referred repeatedly to the "methodical selection" of breeders, and throughout his book emphasizes the slow process of adaptation in response to persistent environmental sources of mortality. It is not enough for some DNAs to be better at replicating than others for some reason at some point in time. In addition to that, it is necessary that some difference in replicating capacity, and its causes, arise repeatedly. An entirely capricious environment, with no recurrent circumstances, could not foster adaptation. The process of adaptation requires that fitness differences hold with some regularity, some recurrence. Adaptation proceeds because of that recurrence, and what we refer to as an adaptation is a distillation, an effect, of that recurrence—a kind of marker to the recurrent circumstances that caused it. This is easiest to appreciate for complex adaptations, such as the eye which could not possibly have evolved unless a particular part of the electromagnetic spectrum was a common feature of the environment for many millions of years *and* organisms that were better at detecting that light also left more offspring (or the DNAs that were better at building eyes also left more descendants). Eyes are surely one effect of light shining steadily on a world of DNA replicators.

But complex adaptations like eyes are just a buildup of many simpler adaptations caused by circumstances that are recurrent in the long term. The point is just as pressing for the slightest of evolutionary changes that arise by natural selection. Natural selection simply cannot lead to adaptation in the absence of recurrence. Every case of adaptation, indeed every particular feature that we might label as an adaptation, is a kind of reference to (or if one prefers, an effect of) the recurrent circumstances in which it evolved. This is so because an adaptation is just an improved capacity for recognition of recurrence, or of responding to recurrence. Certainly eyes function to detect recurrent light, and adaptations to heat are features that work when it is hot. Consider too all those biomolecules that have specific capacities for sticking to certain other kinds of molecules. The best examples of these are the genomic molecules, DNA, and its probable evolutionary precursor, RNA (early in the origin of life and also in some viruses). Not only do these molecules carry information, the coded program for an organism, but also they are inherently capable of recognizing something very like themselves. Every single strand of DNA or RNA is capable of binding to a complementary molecule, a mirror image that is reversed in direction and cast in the alternatives of the G-C and A-T (A-U in RNA) nucleotide pairs that were discovered by Watson and Crick. But beyond nucleic acids we must also consider every protein, every fatty acid, every sugar molecule, indeed every kind of molecule that a cell may synthesize and for which the capacity for synthesis has evolved. This is a large class of molecules, roughly encompassing all those components of a cell or an organism that can be said to have a function. For when we refer to the function of molecule in a biological context, don't we mean that that kind of molecule plays a role in some recurrent context; and don't we mean that that kind of molecule somehow recognizes another, and exerts an effect? Once again we have a conventional idea, this time concerning the function of adaptations, that can be cast in terms of recurrence and its effect in an evolutionary context. Though unconventional, it is perfectly general—and necessary for our present understanding—to see the evolution of function in terms of recurrence and recognition of recurrence.

Recurrence Cycles

While we appreciate that adaptation, and the origins of the earliest replicators, are consequences of recurrence, we must also recognize that once the process of adaptation began, it also became a generator of recurrence. Every instance of adaptation is necessarily manifested in a multiplicity of replicators. Neither the eye, nor any other complex phenotype—and certainly not any DNA capable of giving rise to a complex phenotype—exists as a singularity. Though beginning as just one or a few mutations, adaptations arise by the successful multiplicity of the things that carry them, and ipso facto they cannot happen to one DNA or to one organism. The process of adaptation that occurs today on earth was seeded by the earliest replicating molecules. And from its initial fits and starts, the evolution of life on earth has been a process of the recognition of recurrence, one that focuses on recurrent circumstances and generates in turn, more recurrent circumstances. We can think of ADAPTATION as the name for a kind of two step recurrence cycle in which the first step, recurrence in the environment, leads to the second, which is more discrete recurrence among DNAs. At the center of this process is a turnover of DNAs, a replacement of DNAs with new ones, that though similar also carry a distinct difference in the DNA sequence. At the level of DNAs, the recurrence that caused the change has been distilled into something that is far more distinct, and more recurrent, than the more variable environmental circumstances that caused it.

An adaptation is like a condensation of the commonality among the experiences of the DNAs that are under natural selection, such that the commonality of those experiences becomes manifested as distinct changes in new DNAs (Dawkins 1998, chap. 9). A new adaptation cannot be perfectly suited to all of the circumstances that gave rise to it. Consider one of the better-studied examples of adaptation, at a gene that makes an enzyme, and for which higher or lower temperatures have lead to a new form of the gene that makes an enzyme better suited to those new extremes. There are now many such case studies of adaptation, and in each case the function of the different, related forms of an enzyme have been described in terms of their activity over a range of temperatures (Fields and Somero 1998; Somero and Low 1976; Zavodszky et al. 1998). The optimal temperature, under some measure of activity will be different for one form of an enzyme than for another that works better at a different temperature. But nobody would seriously suppose that each type of the enzyme evolved under just one particular temperature. The adaptation whereby an enzyme has an optimal temperature is an effect of the range of circumstances under which it has evolved, and we may the see the adaptation as a distillation of those circumstances.

As an adaptation arises, the recurrence cycle turns again because of the effects of a distinct DNA change that are again more variable than that DNA change. The effects of a DNA change upon an organism, both in immediate biochemical and physiological term, and in other more disparate aspects of the phenotype, as well as the variation that arises among different organisms that carry that adaptation, will bear some degree of recurrence. But there will also be variation among the phenotypes of organisms that carry the distinct change in DNA. The distilled recurrence among the DNAs gives rise to more variable re-

THE ORIGIN OF NATURAL KINDS 121

currence as it becomes expressed in organisms. Indeed, something very like the indistinct recurrence that leads to successful replicating of the distinct DNAs arises as those DNAs are expressed. We can see that the cycle of recurrence that begins with limited similarity, and that then proceeds to a distillation in the form of many copies of essentially identical discrete DNA changes, begins anew as those DNAs exert their effects and give rise to recurrence that is again more nebulous.

The process of learning, by a nervous system or an immune system, is also at the center of a recurrence cycle. Learning happens when recurrent circumstances give rise to a particular behavior, wherein a key aspect of that behavior is the recognition of instances of the same recurrence that caused the learning. Just as an adaptation is a distilled effect of recurrence, so too is a learned response. And just as an adaptation gives rise to more recurrence among the diversity of phenotypes that carry it, so too does a learned response emerge as recurrent behavior. And just as an adaptation at the level of the DNA is more distinct and invariant than the effects of that adaptation, so too is the physiological basis of a learned response more distinct and invariant than the different manifestations and effects of that behavior.

A Bridge to Language

The question motivating this chapter is whether our typological tendencies, that are revealed as prototype effects, are either fundamental or peripheral aspects of the way the mind forms categories. Under the non-typological, classical, rule-based view, categories are sets of things formed by membership criteria. This is the way that categories seem to be represented in language, and in some ways it would make sense if a classically categorical mind underlay language, both primordially and contemporaneously. Evidence of prototype effects and typological thinking shows that the mind is not simply a classical categorizer. But since language appears semantically classical, and since our brains seem to handle classical categories in many technical contexts, it is reasonable to wonder whether prototype effects and typological thinking are somehow peripheral to the way the mind constructs categories. I don't think that is the case, and the thesis of this chapter is that typological thinking is a fundamental aspect of our minds. The reason is that at the heart of the capacity for language is the capacity to learn, and learning necessarily proceeds as a recurrence cycle.

Unfortunately, we do not know how the capacity for language

evolved. However we do know that 2-year-old humans can do with language what chimpanzees and bonobos cannot do even after years of training. Notwithstanding the possibility that some apes can gain some limited capacity for syntax and for using symbols to refer to things (as we use words) if taught very extensively from the time they are very young (Savage-Rumbaugh et al. 1998), the language of humans is an exceptional behavior. But there is an interesting irony in the history of our understanding of this exceptional capacity. The irony lies in the common view, over the past 40 years, that the capacity for language is innate and is *not* an adaptation. Noam Chomsky convinced nearly everyone of the human innate capacity for language, and he also convinced many that this ability was best seen as a byproduct, an epiphenomena, of other adaptations that shaped the brain. In recent years, this view has suffered under attacks from philosophers and linguists (Dennett 1995, chap. 13; Newmeyer 1998; Pinker 1994; Pinker and Bloom 1990). These assaults have made much of those points where Chomsky's arguments rely on a misunderstanding of natural selection, and even more of the fact that strong evidence of complicated function is also strong evidence of design by natural selection. The function of language is all that we use it for (and that is a great deal), and the evidence that language is complex lies in its multifaceted intricacies. (Witness how significant sectors of multiple fields—physiology, psychology, linguistics, and anthropology—are devoted to the study of different aspects of language.) Note too that when we speak of design in the context of evolution, then the process by which design arises is, of course, natural selection. At this time, I think the most interesting questions regarding adaptation and language are in the details, and over just what aspects of language are best seen as adaptations (Deacon 1997; Hurford et al. 1998).

But regardless of debate on language as an adaptation, we certainly understand that the acquisition of a particular language happens via learning. Also, I think it is fair to say that psychologists and biologists would accept that the capacity to learn is an adaptation, or rather like all complex aspects of phenotypes, it is the accumulation of many adaptations. Now consider the way that learning happens as part of the acquisition of a particular language. As with any learning, the learning of a new word is a function of perceived recurrence in the environment. The facility for learning a word, especially in children, is so efficient that the impression of similarity may be limited to two cases, the child's first exposure to the word, and the circumstances that inspire the child's first usage of the word. But like an adaptation in DNA that is caused by recurrence in the environment, so too is the learning of a word inspired by recurrence in the environment. In the case of adaptation, the mechanism is natural selection, whereas in the case of a new word it is learning. Also like adaptation, the learning of a word does not require perfect recurrence; rather the new word is a distillation of perceived commonality among the recurrent circumstances for which the word is used as a symbol. People learn words from reading and from hearing others, and our exposure to new words comes as they are paired with their referents. But whereas a new word, as text or a noise from the mouth, may be distinctly recurrent, the circumstances for which it is used are far more variable and less recurrent. When a child learns DOG by hearing it repeatedly in a range of partly recurrent contexts, she may hear a distinct highly recurrent set of phonemes, but what those sounds are paired to are likely to be a highly diverse set of animals and pictures. Word learning and usage are processes that necessarily play out in the midst of a great deal of vagueness. As Quine noted in *Word & Object*, "vagueness is a natural consequence of the basic mechanism of word learning. The penumbral objects of a vague term are the objects whose similarity to ones for which the verbal response has been rewarded is relatively slight" (1960, 125).

The learning of a word is also, like other learned behaviors and like an adaptation, at the center of a recurrence cycle. Variable recurrent circumstances are distilled by learning into a distinct word, and when that distinct word is expressed it unleashes a variety of effects. When the word is spoken, or written, it is highly recurrent. But it occurs in a wide variety of sentences and many different instances of it may be perceived by different people. In short, the effects generated by different instances of a spoken word are far more variable than is that word. As with adaptation, word learning is at the heart of a recurrence cycle: from variable and indistinct recurrence; to a discrete distillation in the form of a word; and back to variable and indistinct recurrence among the effects of that word.

Now let us ask: What, in the mind, represents a learned behavior? That question lies at the heart of much of the modern cognitive sciences, and I don't presume to provide anything close to a full answer. But if we follow the argument, then there is a connection between the adaptations of DNAs and the learning of behaviors. Recall from earlier in the chapter that an adaptation occurs when a discrete change in DNA that aids in replication arises as a distillation of recurrence in the environment: It is a reference to that recurrence, however vague that recurrence may be. Furthermore, since that recurrence is vague, the

DNA change refers more to some instances than to others (i.e., it was caused more by some instances than others). Now consider that the capacity to learn arose as an adaptation, and that each instance of learning is also one in which recurrence is distilled into a behavior. How is that distillation, the learned information, likely to be represented? One idea, that we may borrow from theories on category acquisition (chapter 4) is that a category is represented as a set of rules. The major alternative is that there is some more prototypical representation, or at least something that gives rise to prototype effects (fuzzy categories, with good and bad examples). We know one way that learning really does happen in one situation at the molecular level, and that is within the workings of mammalian immune systems. There the immune system learns to recognize recurrent antigens by devising specific molecules that match the antigens. In short the immune system does not use a rule. But neither does it really devise a prototype of the antigens. What it does is to devise some things, antibodies, that will stick to antigens (some more than others). Such a system will, and does, generate a range of immune responses and specificities: Antibodies exert a range of specificities with strongest reactions for those antigens most similar to the ones that caused the production of the antibodies (which are just proteins after all). In short, while the immune system does not employ literal prototypes of antigens, it does employ sticky protein molecules that do exhibit prototype effects. What would we expect of the learning systems that evolved in simple organisms? Would we expect something like a rule-based system? Or would we expect something that mimics the adaptation process that gives gave rise to the learning capacity, something like what we see with mammalian immune systems, a molecular recognition system that exerts prototype effects? I have not proven it, but I suspect that the latter kind of system is what would evolve.

The argument that is laid out here is an explanation of prototype effects in learned behaviors. In the next section I consider more directly the evolution of the capacity to learn words, as symbols. These arguments are reduced explanations of how such behaviors may have evolved, and thus of what manner of neurological basis might exist for such behaviors. They are general, and vague, but in that general, vague way they do predict prototype effects.

Symbols and Learning

What is special about the learning of a word, as compared to other learned behaviors, is the presence of a word, a symbol, in the midst of the recurrence cycle. In between the perception of recurrence in the environment, and the actual behavior (the use of a word) is the mental acquisition and deployment of a symbol. Probably most learned human behaviors, and certainly all behaviors learned by other organisms, arise without the acquisition of a symbol. A symbol is an arbitrary entity that stands for something else, and it exists merely by social convention, because of those who are prepared to interpret it. A nation's flag is a symbol, and so is each word in this sentence. Deacon (1997) has argued that the key adaptation—the one that sets human minds, and human language, apart from apes—was the acquisition of a general capacity for using symbols. The basic argument draws much from the work of Charles Sanders Pierce on different modes of reference, particularly on the distinction between indexical and symbolic reference.

Indexical reference is when one thing indicates another because of some direct causal connection. For example, the direction pointed to by a weathervane is an indexical reference to the direction the wind is coming from. Of course, this use of REFERENCE does not imply anything mental, but rather just some kind of cause and effect connection. Thus the use of REFERENCE only makes sense when we consider the interpretation, or recognition, that one thing is associated with another. Here is where learning comes in. Under a basic operational definition of LEARNING, a change in behavior as a result of a perceived association of one thing with another, learning is a process of recognition of recurrent correlations. Of course, RECOGNITION in this sense is not intended to imply consciousness, but just the establishment of whatever physiological basis exists for the connection between sensations and the learned behavior. It is in this sense of invoking the observer's interpretation that a weathervane can refer to the wind. So too, for flatworms in a petri dish, can a chemical refer to food, if that is what they have learned. Deacon also gives the example of learning to interpret the smell of smoke as a sign of nearby flames, which can happen simply by repeated co-occurrence of smoke and flames. In short, associative learning whether by a person, or a far simpler nervous system, is recognition of recurrent associations.

The advance that humans have made is from indexical reference to symbolic reference. They have not given up on indexical references, but on top of that they have gained a tremendous facility for recognizing interconnections among indexical references. Suppose a simple mind comes to be fairly full up of knowledge of indexical relationships, of all the sorts of correlations that have been impressed on that mind, and that now exist as learned responses. What would happen then is that some of the indexical relationships that are represented in the brain (however they may occur) are recurrent of one another. And just as the brain learned of some indexical relationships, it may also come to learn that many indexical relationships are like one another. Perhaps it is not too difficult to envision that a network of interconnections among learned indexical relationships might build up in a very complex web. The more learning occurs, and the more capacities to recognize indexical references that the brain has, then also the more potential for learning that one such connection is like another, and so on and so on. Deacon then supposes that to organize the "knowledge" of many associations and of similarities among associations, that what happens is a sort of mental recoding. Rather than persist in a method where each indexical association is connected to every other one that resembles it, a mnemonic method emerges that contains the commonalities of the different connections between indexical associations. It was this switch, from having many indexical associations (many interconnected and many similar to one another) to a method for organizing those commonalities, that underlies symbolic reference. A symbol, or its mental representation, is not an index that was caused by something. Rather, it is something that stands for a recurrent pattern among indices. Precisely what that symbol is does not matter, so long as it is somehow wired to knowledge of recurrence. Nor are symbolic references constrained in space and time. In any particular case where I smell smoke and interpret fire, my interpretation is in the same moment of the smoke and the fire. But the potential interpretations of CAT include future or past cats, imagined cats, and things that look like cats, all depending on what other word symbols surround CAT. Another key point is that, like the web of indexical relationships that give rise to symbolic ones, symbols do not stand in isolation but serve to refer to things by their interconnections to other symbols. A single symbol would serve nothing, but as a part of a web of symbols they can represent and generalize a more tangled web of indexical relationships.

The details of Deacon's argument are lacking, and it is a fairly abstract hypothesis. He supports it well with insights from the ways that chimpanzees and bonobos learn the rudiments of language. But one of the best features of Deacon's story is its consistency with respect to learning and the capacity to learn. At each stage in the rise of symbolic reference, Deacon proposes a process whereby recurrence is recognized. In this light, the origins of symbolic reference, and thus of one of the key aspects of language (though far from the only one), can be seen as the piling up of increasing adaptation and improvement in the capacity to learn. If Deacon is correct, then we can see that the capacity for using a symbol for a category is a behavior with an evolutionary precursor that is simple associative learning—the capacity for interpretation of indexical reference.

Though language is far more than symbolic reference, having an understanding of how symbolic reference might have evolved does help us in our consideration of the role of typological thinking in our use of categories. Whatever categories are in the mind, we do represent them with symbols (words in speech and text). But if the capacity for symbolic reference is a kind of fancy capacity to learn, then we would expect some similarity between the way the mind holds knowledge of what a symbol can stand for, and the way that a learned indexical relationship is held in the brain. Recall that a simple learned behavior is like an adaptation; it is a distillation of recurrence: whatever is contained in the neurons that hold that behavior, it is a condensed representation of some commonality among the recurrent circumstances that caused it. The neural basis of the behavior is an inherently typological reference to the recurrence that caused it. Among the possible circumstances that could evoke the behavior, some are quintessential and some are peripheral.

If the evolution of language was indeed an extension of the earlier evolution of the capacity to learn, then we can expect, whatever the mental basis of the typology of learned indexical references, that the typological aspect has not been entirely discarded with the rise of symbolic reference. What we see today is evidence of prototype effects, and typological thinking underlying the use of human categories. Given the apparent connections between our capacity to have categories, and basic aspects of learning, our modern day typological thinking and prototype effects are probably manifestations of very basic, and very old, ways that the brain contains knowledge.

The Theory of Life and the Origin of Categories

The purpose of all this is to complete the explanation of the species problem. In the previous chapter I explained how the species problem is caused both by our insistence upon categories as species proxies and by our demand to understand species as entities. The more we understand species as entities, the more we see that they do not correspond to our categories; yet for some reason that understanding does not dispel our doggedness on the categories. That is the explanation, but it still leaves some important questions. Even if correct, the explanation should be seen by scientists as a proximal one. We must still wonder: "Why does our categorical imperative lie so deep?" and "Why are our categories of organisms partly typological?" If the theory presented in this chapter does come to be seen as answering these biological questions, or if it does not and some other theory does, then we will have come near to closing the loop and providing a biological explanation of the species problem. It may not be of the variety of biological resolution that many have envisioned for the species problem—it certainly is not a mere species concept—but at least we are still dealing in the terms of the theory of life. Recall that at its core, the argument in this chapter is based upon the idea of a recurrence cycle; and recall too that the process of adaptation of DNA replicators is at the center of a recurrence cycle.

The theory of the evolution of categories that is presented in this chapter is fairly sketchy. It is verbal and at points the argument relies upon an omniscient perspective of no particular validity except that it seems both workable and necessary to make the case. Finally, the last few steps draw heavily from Deacon and are quite an abstract mix, seemingly much more philosophical than psychological or biological. But the theory does fit its purpose. It explains, in a qualitative way, our strong capacity for creating typological categories and it does so by building on the theory of life.

Part II Conclusions

The previous six chapters have laid out the primary argument of this book. The species problem explanation is, at least in some ways, not very complicated. The basic claim, which is not new, is that humans have conflicting motivations when we consider species. However, the explanation of the causes of those motivations, and of just why they are in conflict, has required a bringing together of ideas and discoveries from several different fields. Our confluence of those different fields is the subject of categories. Regardless of how species exist in nature— outside of human minds and communication—their syntactical position in human language, and their representation with human minds, is that of categories (chapters 4 and 5). But as much as we refer to and think of species in a categorical way, we also have another, quite different and more modern way of thinking of species. Since Darwin, a dominant theme of debates over the nature of species concerns their existence as evolutionary entities in the natural world. The precise nature of such entities is hard to perceive through our species problem haze, but it is not difficult to build a theory of them. In fact, some straightforward extensions to the theory of life that was described in chapter 3 lead to the idea of evolutionary groups, which has a great deal in common with many ideas of species as evolutionary entities (chapters 6 and 7). The trouble is that these sorts of entities have an inherent and poor correspondence with our mental, categorical representations of species. Finally, since we are generally unaware of our basic mental dependence upon species as categories, and of the poor fit these categories must have with real evolutionary groups, we suffer the species problem (chapter 8). The final chapter of this second part of the book has addressed the evolutionary question of why human mental categories are structured as they are and why they have the largely typological component that is not well suited to an appreciation of real species.

In a general way, the understanding of the causes of the species problem, by way of categories, seems to fit the argument and the behaviors that motivated the search for that explanation (as outlined in part I). The argument, arising as it does from biologists' measurement errors regarding species and from the ways that biologists persist in making those errors despite their knowledge of them, called for a search for a psychological explanation. Now that we have that kind of explanation, what remains is to return to the ways that biologists study species and to assess the day-to-day phenomenology of the species problem in light of the explanation (chapters 11 and 12). Following this, we return in the final two chapters to the basic question of the nature of real species, and of the nature of species taxa, and of how we should study and refer to them.

However, before considering the work of biologists who study species, we need first to consider the evolutionary processes of diversification. So far the theory of evolutionary groups has left fairly implicit the idea that groups may diverge from one another. This minimal description of divergence does not do justice to the complexities of evolutionary diversity, though it can be extended to do so. In practice, on a day-to-day level, the study of closely related species is embedded within the study of more divergent kinds of organisms. Thus a prerequisite for examining the work of systematists and other evolutionary biologists is to extend the theory of evolutionary groups to include a more complete description of diversification. This is undertaken in the next chapter.

PART III

Living with the Species Problem

TEN

PHYLOGENY

The title of this chapter is a word that is used frequently by biologists to refer to the origin (i.e., genesis) of phyla (traditionally, kinds of organisms). On the basis of its etymology, the word could be used for any or all aspects of the origins of biological diversity, but in modern parlance, it usually refers to a branching history of evolutionarily related groups of organisms. The use of branching diagrams to represent relationships among kinds of organisms predates Darwin, but one of the greatest byproducts of Darwin's discoveries is the theoretical support they provide for the use of tree-shaped diagrams. Darwin wrote

> The affinities of all the beings of the same class have sometimes been represented by a great tree. I believe this simile largely speaks the truth. The green and budding twigs may represent existing species; and those produced during each former year may represent the long succession of extinct species. At each period of growth all the growing twigs have tried to branch out on all sides, and to overtop and kill the surrounding twigs and branches, in the same manner as species and groups of species have tried to overmaster other species in the great battle for life. The limbs divided into great branches, and these into lesser and lesser branches, were themselves once, when the tree was small, budding twigs; and this connexion of the former and present buds by ramifying branches may well represent the classification of all extinct and living species in groups subordinate to groups. ([1859] 1964, 129)

Today, most everyone with some exposure to the biological sciences knows about evolutionary trees, and takes their use for granted. But are they really a good way to depict evolutionary history? The answer is far more complicated than many biologists suppose or would wish.

Tree forms are certainly well justified for describing the relationships among ancestors and descendants of non-recombining homologous DNAs (see Figs. 6.1 and 6.2). In this context, the theoretical correspondence between the portions of a diagram and historical entities is clear. A branch point, or node, represents an ancestral DNA that underwent replication. And, to belabor the point, we know that this occurs by the laying down of a new strand of DNA upon each strand of the original double helix (Fig. 6.1A), and that this is carried out by the DNA polymerase enzyme. Furthermore, a node can represent only one of a highly specific subset of all the replication events that occurred in the history of sample of DNAs. For a replication event to be represented as a node, it must have been one for which both daughter DNAs were ultimately the ancestors of DNA molecules at the tips of the tree. These few replication events are, of course, but a small minority of those that occurred in the history of a sample of DNAs. Most replication events in the history of a sample of DNAs are not represented in a tree diagram because they do not meet the stringent criteria for inclusion in the tree (i.e., both daughter sequences must have been ancestors of sampled DNAs for a replication event to be represented by a node). The actual branches of the tree correspond to the persistence of a DNA through time. That persistence includes simple physical persistence, but also includes numerous cases of replication from ancestor to descendant, when it is the information in the sequence that persists. These hidden replication events are not represented as such in the tree (nor can they be directly discerned or counted by tree estimation methods). In short, the parts of a tree graphic can correspond quite well to the kinds of historical processes that we envision as possible for a sample of homologous (nonrecombining) DNAs.

In contrast to the case for DNAs, Darwin's justification for a tree diagram for species or taxa is not so simple or as plainly justified. Though compelling as a metaphor, the tree model for taxa comes with only a crude explanation of how one taxon can split, or replicate. However, it is not difficult to embellish Darwin's metaphor, and include with it a model of how one evolutionary group may split to become two. Let us carry forward the simple no-recombination replicator model, to see how it might reveal the circumstances wherein tree models for evolutionary groups are, or are not, likely to be fitting descriptors of evolutionary history. To do so, it will help if we employ a common graphical device for considering the kinds of relationships that can arise between DNA trees and the trees for evolutionary groups of organisms that contain those DNAs. Like many authors, I've often found it use-

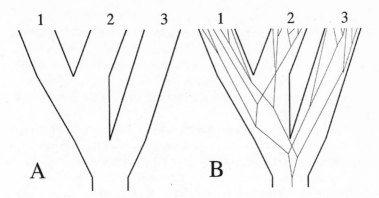

Fig. 10.1. (A) Two recent splitting events that gave rise to three evolutionary groups. (B) The DNA tree of a sample of DNAs drawn from each of those evolutionary groups.

ful to draw a "skinny" DNA tree (as in Fig. 6.2) within the confines of a wider "fat" tree that represents the history of several groups of organisms (Hey 1994). In Fig. 10.1A, each pair of widely spaced parallel lines is a branch that represents the persistence of an evolutionary group of DNAs, in the sense that such groups were described in chapter 6. The junctions of these wide branches correspond to the splitting of one ancestral evolutionary group into two descendant groups. The presumption that these splits are instantaneous is an assumption for this example. This is the kind of history that might occur if one evolutionary group suddenly became split by geography. Figure 10.1B repeats the figure but drawn within it is a random branching DNA tree, constrained only so that no node can leave descendants in multiple evolutionary groups if that node occurred at a time after the splitting of those evolutionary groups (i.e., skinny gene tree branches cannot cross fat branch boundaries that represent evolutionary groups).

Now consider a sample of just one DNA, a single branch tip, drawn one from each evolutionary group, and compare the shape of the branching pattern (the topology) of the skinny DNA tree to the shape of the larger fat tree. Note that depending on which particular DNAs are considered, the branching pattern changes, and it will often not match the topology for the fat tree. The reason is just that the common ancestor for many of the DNAs that currently exist in one evolutionary group actually existed prior to the major splits in the fat tree representing the splitting of evolutionary groups. The potential for these incongruencies is well known (e.g., Tajima 1983), and to those concerned with getting the correct tree for species using a few DNAs from them, it is called the gene-tree-species-tree problem. Indeed, but for the basic constraint that DNAs from different groups must have branch points earlier than the split of their groups, there is no necessary relationship between the branching patterns of DNAs and for their corresponding evolutionary groups.

However the forces that cause evolutionary groups will also cause a tendency for DNA branching patterns to resemble the history of the evolutionary groups from which those DNAs are sampled. As turnover occurs within evolutionary groups, by genetic drift and beneficial mutations, so to will the pattern of ancestry of DNA trees, within groups, shift forward in time (Fig. 6.3). Thus, if the splitting of an evolutionary group occurred a long time ago with respect to the rate of turnover, *and* if these splitting events were separated from one another by considerable time, then there is the expectation that the DNA trees will have the same topology as those for the evolutionary groups (Pamilo and Nei 1988). However, if the splitting of evolutionary groups was recent, or multiple such events occurred near each other in time, then the history of just single DNAs, one from each evolutionary group, may not match that of the evolutionary groups. One thing that can be done is to use a phylogenetic model that recognizes that individual DNAs may not match the branching pattern of the evolutionary group. Such methods rely upon the amount of DNA sequence variation, within and between closely related groups, to infer the branching history of the groups (Wakeley and Hey 1997). Methods like these generally rely on the assumptions that evolutionary groups have sharp boundaries, and that there is an absence of population subdivisions within groups, and that the splitting of evolutionary groups happened instantaneously. Under these assumptions, the gene-tree-species-tree problem is somewhat tractable, even if splitting events were recent or close to each other in time. But note that these assumptions are the equivalent of assuming that evolutionary groups have truly branching phylogenies.

When we must consider recombination, which is most of the time, the correspondence between DNA histories and taxon histories gets more complicated. If recombination has occurred among the ancestors of a sample of DNAs under consideration, then those DNAs cannot literally have a single branching history. If multiple DNAs are drawn from each taxon, it is sometimes possible to see the evidence of these recombination events from the pattern of variation (Hudson and Kaplan 1985), and actual data sets may show these patterns and reveal the historical recombination events (Kliman and Hey 1993). When just one DNA is drawn from each taxon, as is typically done for many phylogenetic studies, then recombination may not be revealed. Nevertheless, it may still have occurred and any simple tree model that is imposed on the data may be in error for assuming that recombination has not been present.

Recombination is also a major issue when considering DNAs drawn from more than one region of the genome. Because genomes are large and recombinogenic and because, for most organisms, the genome is divided up into multiple chromosomes, a study that includes DNAs from two or more regions of a genome will often be considering multiple unlinked histories. The histories are unlinked in a literal sense, as the particular DNAs that have been sampled from multiple genomic regions have not been consistently attached to one another throughout their history. These histories must still have been shaped by those evolutionary forces that acted upon the genome as a whole, but their histories can also differ by chance and by natural selection that acts differently on different parts of the genome. Thus, when recombination has been happening, we should consider not only that a DNA history may not correspond to that for the evolutionary groups, but also that different portions of the genome will have no necessary correspondence to each other, save that all must fit within the history for the organisms that have been sampled. Different parts of the genome of recombinogenic organisms must have different histories. Though they may be similar, we must recognize that because of recombination the historical events in the histories of two different genes are necessarily different ones.

As with nonrecombining DNAs, there is an expectation for the multiple trees of a sample of recombining DNAs that with time following the splitting of evolutionary groups, all the trees will tend to resemble, in shape, the history of the evolutionary groups from which they came. This is a powerful prediction, and it is a primary motivating idea behind modern DNA based methods of studying recent historical speciation events. Whereas any one portion of the genome may not well reflect history, a study that includes many genes can reveal not only the species phylogeny but also much about the sizes of ancestral species and the times of splitting (Hey 1994; Wakeley and Hey 1998).

Perhaps the best example to help make these points comes from the study of human evolutionary history, and of our evolutionary relationship to other apes. Let us suppose for the sake of argument that humans, and chimpanzees and gorillas are each distinct species in the sense that the organisms within each of them share in evolutionary processes (i.e., they are each evolutionary groups). This supposition is probably not accurate for gorillas (Garner and Ryder 1996) and is clearly not so for chimpanzees, which are known to have two quite distinct species (common and pygmy chimpanzees) as well as multiple partly distinct populations within species (Kaessmann et al. 1999; Morin et al. 1992). But it is still possible that each of the recent ancestors of each these groups were distinct evolutionary groups, and for the moment let us suppose that they were. Now consider the data compiled by Ruvolo representing sequences from each of these groups, for each of 14 different genes. A majority of the genes that were examined (11 out of 14) revealed estimates of history in which the DNAs from humans and chimpanzees appear to be more closely related to each other (Ruvolo 1997). The remainder of the genes suggested histories in which either the human and gorilla DNAs were most closely related, or the chimpanzee and gorilla DNA were the most closely related. If we are assuming that the true history was of distinct evolutionary groups and clean splits, then what are we to make of these disparities in gene tree topologies? If fact the data are entirely consistent with such a model, as exemplified in Fig. 10.2A. If there were two splitting events of ancestral evolutionary groups, and they occurred near one another in time, then it is quite reasonable to have DNAs from some genes show one kind of history while other DNAs show different histories. On balance, the data support humans and chimpanzees as each other's most closely related species (Ruvolo 1997). Also, it is important to keep in mind that this is a statistical statement based on the pattern found at a majority of the genes studied, and there is a chance that it is incorrect.

Things are even more complicated when we consider that evolutionary groups need not have sharp boundaries. It is possible that the patterns of DNA trees that we find today among apes, including humans, may have been caused by a history in which ancestral evolutionary groups underwent fission by a gradual process during which gene exchange occurred (Fig. 10.2B). This is not a model of clean splits and distinct groups, and it is not a model that is easily ruled out. Finally, it is a model in which the true history of present-day taxa is not one with simple, or discrete, branching events.

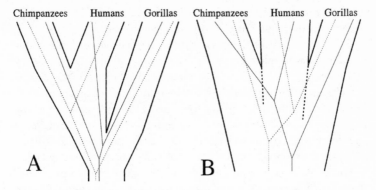

Fig. 10.2. (A) The history of three taxa, each corresponding to an
evolutionary group. The history of the groups is shown as a fat tree with two
distinct splitting events. Within the evolutionary group histories lie the
histories of two different samples of DNAs (straight and dotted thin lines),
drawn from two independent genes. (B) A history in which three
evolutionary groups emerged from a single ancestor within which partial
boundaries (large dotted lines) arose prior to the complete separation of the
groups. These partial boundaries allow the limited movement of genes
between evolutionary groups. The histories of two DNA samples, from two
different genes, are shown within the fat tree history.

The Use of Tree Diagrams

These descriptions of trees, both for DNAs and for evolutionary
groups, follow directly from the theory of replicators and evolutionary
groups as outlined in the previous chapters. For the most part, they are
not novel ideas, having been expressed in different parts by many authors in recent years. However, they do differ from the traditional way
in which evolutionary tree diagrams are considered. In that tradition, a
tree representation of history has taxa (typically of species rank, but
sometimes they are genera, or families, or from even higher ranks) at
the tips of the branches, and hypothetical ancestral species at the nodes.
The utility of evolutionary trees for taxa was appreciated well back in
the 19th century, and the validity of such trees has been a ubiquitous
assumption at least since Darwin's theoretical justification for them. In
contrast, the fact that DNA must have a branching history, at least over
nonrecombining stretches, is a more modern discovery. The modern
addendum to the evolutionary tree tradition is of a switch, away from

trees that reflect the history of species, to trees that are more literal representations of samples of DNAs from those species.

The good news, for those who would use estimates of DNA histories as proxies for phylogenetic estimates, is of an expected tendency toward correspondence. We can indeed expect there to be a large branching component to the evolutionary histories of organisms. The ultimate reason for this is that DNAs really do undergo a branching process as they replicate and leave descendants. Replication (branching) and the pruning of trees, by genetic drift and natural selection within evolutionary groups, leads to patterns of recent ancestry among DNAs within evolutionary groups. In contrast, the gene trees from samples that include distinctly separate evolutionary groups can be quite deep, arbitrarily deep, as separate groups can diverge indefinitely from one another by the forces of evolution. Though recombination allows different genes to have different histories, and natural selection may actively force them to do so, the forces that generate evolutionary groups will also contribute to a tendency for different genes to have similar evolutionary histories. The tendency for correspondence, between gene trees and the histories of evolutionary groups is stronger when evolutionary groups are distinct, and when they split over short time periods, and when multiple splitting events are separated by long time intervals. To the extent that these processes hold, then a study of just one or a few genes may be used to estimate the history of the groups from which the samples were drawn.

The bad news—or to be less negative, the complicating news—is that evolutionary groups, and the splitting of evolutionary groups need not have those properties that tend to cause simple branching histories. Certainly our simple model of evolutionary groups reveals no necessity for distinct groups or sudden complete splitting events. We also know from many studies on species and speciation events that there are good reasons and frequent causes for trees of DNA histories that do not fit those for the species from which they came. The most egregious exceptions to this view are those species that have been formed via hybridization events between organisms of different species (Arnold 1997). This mode of origination of species is not a splitting mechanism at all, but a process of joining. No less un-treelike are those cases where the genomes of organisms clearly have ancestry in quite different historical groups. It has become clear that horizontal transfer among prokaryotes is a common process (Jain et al. 1999; Lawrence and Ochman 1998), and it appears that eukaryotes arose as a mosaic of genes from groups whose descendants we now tend to think of as two divergent and ancient groups, the eubacteria and the archaebacteria (Gupta and Golding 1996; Gupta and Singh 1994; Lopez-Garc and Moreira 1999). There are also many more commonplace, and evolutionarily recent, contexts where a branching model for the history of phyla is incorrect. These are the numerous cases where we know that different parts of the genome of related species have quite different gene trees and where it appears that gene flow as been occurring between species for some genes and not others (see chapter 7). In these cases, we find that at least one part of the genome has a branching history that is simply in conflict with that of the remainder. Such cases are like that depicted in Fig. 10.2B, where splitting has occurred gradually with periods of gene exchange. If we accept that these kinds of histories occur, then we must logically also accept that some species do not literally have a branching history. If a species is a real entity, then necessarily it has just one history (even if the entity is not distinct), but that history can be a mosaic of processes and need not be a simple branching one. How common are such mosaic genomes? The short answer is that nobody knows. But one lesson of recent years is that the closer we look, and the more that the histories of different portions of genomes are studied, the more evidence we find for mosaicism. These lessons also help us to recall that most awkward lineament of the species problem: the traces of knowledge that biologists suppress. The tree model is invoked by biologists without question in the vast majority of phylogenetic research, and yet we can be sure that virtually all biologists know that speciation is often gradual and that real evolutionary units in nature may often not be distinct.

But should we discard tree models for species? The answer is "yes" for those species that form by the joining of other species, and for organisms that regularly exchange genes among widely divergent forms. It remains to be seen what proportion of biological diversity cannot have its history approximated with a tree diagram. But for many eukaryotes, the answer is clearly "no," as there are two very good reasons, one empirical and the other theoretical, why tree models are a good, practical approximation in many cases. The empirical justification is just that the histories of most genes, particularly those for DNAs drawn from distinctly different species, have been found to be similar. There are not yet many cases where the same sets of divergent taxa have been studied with multiple genes, but those studies that have been done typically find a large amount of congruence among estimated gene histories (see, e.g., Cronquist 1997). We have seen this tendency for most of the species in the Drosophila species groups we studied (Hey and Kli-

man 1993; Hilton and Hey 1997; Wang et al. 1997). Also DNA tree estimates for well recognized higher taxa are *mostly* congruent with tree
estimates that had previously been made using morphological data (see,
e.g., Goodman et al. 1994; Russo et al. 1995). There is actually quite a
good theoretical justification for why DNA trees may yield tree estimates that are approximately correct for their sampled taxa, even if the
histories of those taxa include indistinct evolutionary groups and slow
splitting events. The reason is that, even though DNAs may move between different evolutionary groups and so cause them to be partly indistinct, such movements will still tend to be between closely related
groups. In short, the practical effect of such movements on trees estimated for divergent taxa may be quite small even if such DNA movements are common. Branching diagram estimates of phylogeny have a
kind of robustness because as long as gene exchange between groups
is a declining function of disparity, trees drawn on the basis of different genes will be increasingly similar for more-divergent groups. Tree
diagrams will often be a very poor approximation of history for closely
related species, but for more divergent groups, we expect them to be
more useful. Of course, this is not always the case, as shown by one of
the largest taxa, *Eukaryota*, which appears to have a highly complex,
nonbranching historical relationship to diverse bacteria (Gupta and
Golding 1996; Gupta and Singh 1994; Lopez-Garc and Moreira 1999).

Phylogenetic Taxa

Perhaps the greatest uncertainty that arises in the context of tree models is how to devise taxa. Recall that a taxon is a named group of organisms. Everyone makes regular use of many taxa: PEOPLE, DOGS, and
PLANTS are all labels for commonly used taxa. Biologists also make use
of more formal taxa that carry formal, scientific names and for which
the names have been deliberately placed in a hierarchical classification
scheme. Historically those hierarchies roughly reflect patterns of similarity recognized by systematists. But, increasingly, biologists are interested in having these hierarchies of categories reflect evolutionary history—not just approximately, but via a precise matching of our best
estimates of evolutionary trees with the names and the hierarchies of
taxa. Two question arise: Should we identify a taxon in such a way that
it matches up with an evolutionary tree?; and if so, what is the best way
to do that? Increasingly, the answer to the first question, by systematists,
is "yes." To the second question, an attractive idea is that a taxon should

correspond to a clade (meaning a node and all of its descendants) on an evolutionary tree (Hennig 1966; Wiley 1981). A taxon is said to be monophyletic if it includes all of the descendants of a common ancestral species, and thus corresponds to a clade on a tree. In contrast to monophyletic taxa, paraphyletic taxa do not include all of the descendants of a node on a tree. In other words, if the ancestral node is identified for all of the organisms within a taxon, and if that node is found to have other descendants that are not in that taxon, then that taxon is paraphyletic. The appeal of a monophyletic taxon is the idea that all the organisms in the taxon share more in evolutionary history than they do with other organisms not included within that taxon. This basic idea also translates into a wealth of useful ways that a monophyletic taxon can be used to study the evolution of traits that are shared by its members (Harvey and Pagel 1991; Hennig 1966; Wiley 1981).

As useful as the concept and assessments of monophyly may be in some taxonomic contexts, they have little place in our understanding or assessments of species. If species are considered to be real entities as envisioned by most biologists, or as the evolutionary groups that I have described, then we may give them names, and we may try to place them on a branching model of the history of life. But real species that are recognized in this way are fundamentally different from more inclusive "higher" taxa. They are different because they are identified on the basis of shared evolutionary processes that go on within them. More-inclusive taxa do not share in such things, and so nor do they exist in this way. They are categories that are devised on the basis of patterns of similarities or of shared traits. Occasionally, some authors have envisioned species as monophyletic groups of related organisms or populations (de Queiroz and Donoghue 1988; Donoghue 1985; Mishler and Donoghue 1982; Mishler and Theriot 2000; Rosen 1979). Of course, the organisms that constitute evolutionary groups have no necessity (nor even a tendency, if they are recombinogenic) to have shared histories that could be described as monophyletic. Indeed a fair portion of the recent history of species concept debates in systematics has been realizing the futility of monophyly as a criteria for devising taxa that could correspond well to many ideas of species, such as under the BSC or to the sorts of real evolutionary groups that I have described here (Avise and Wollenberg 1997; Crisp and Chandler 1996; Doyle 1992; Vrana and Wheeler 1992).

Consider what it might mean to construct an evolutionary tree, as a model of history, and to envision current real species at the tips of that tree. The terminal branch leading to a particular tip would supposedly be some kind of reference to, or model of, the recent persistence of a species. But how does a species end up on the tip of a tree drawn by a biologist? If the data that is the basis of the tree comes from just one organism or just one DNA, from each of several taxa, then each organism or DNA is a proxy for one entire species. Alternatively, multiple organisms are used for each species, or multiple copies of a gene (or multiple copies from each of multiple genes), then it is not at all clear how the species should be represented by the tree. Trees based on multiple samples from each of several closely related species typically do not suggest monophyly for those species (Crisp and Chandler 1996). Figure 10.1B is an example of the kinds of paraphyly that can occur among multiple DNAs considered from each of several closely related taxa. It can also happen that a species taxon may be monophyletic at many genes but not at some. Consider the case of *Homo sapiens*. The last common ancestral species for humans and apes is thought to have lived 5 million years ago, or more (Takahata and Satta 1997). Humans all share an abundance of unique derived traits, and almost all phylogenetic trees built with humans, and other primates, would show all the humans to comprise a monophyletic group. But this is true only of *almost* all phylogenetic trees estimated from DNAs. At the major histocompatibility complex loci, humans have a paraphyletic history, and many gene copies share more recent ancestry with those from other apes, before they do with other gene copies in humans (Gyllensten et al. 1990; Klein et al. 1993).

In summary then, tree diagrams are useful for approximating the evolutionary relationships among taxa. However they do not apply in a simple way to the recent histories of evolutionary groups. Such groups can only crudely be represented by a single branch to a single organism or DNA, and if multiple samples are used, then evolutionary groups construed with tree diagrams tend to not be monophyletic.

The topics of taxa and evolutionary trees will come up several times in the next few chapters, as we consider the ways that that biologists study organismal diversity. In particular, the next chapter brings the species problem explanation, and the understanding of phylogeny that has been developed in this chapter, to help in consideration of the major biological subdiscipline of systematics.

ELEVEN

SYSTEMATICS

The species problem does not burden all biologists equally. The primary sufferers are those who study species and must deal with trying to connect our categories of organisms to real evolutionary groups. These issues are especially acute for many in the subdiscipline of systematics. Before Darwin, and for quite a long time after as well, systematics was concerned simply with the classification of organisms. In the absence of biological principals, the field was a "specialized, rather narrow branch of biology, on the whole empirical and lacking in unifying principles" (Huxley 1940, 1), more like art than science (Stevens 1990). But as Huxley foresaw, it has grown with increasing knowledge of evolutionary processes. Today, systematists are concerned not only with devising categories of organisms and with the placement of organisms within those categories (identification), but they are also devoted to figuring out the evolutionary history of kinds of organisms. The two tasks, categorizing and history estimation, seem on the surface to be a natural fit for one another. In both, systematists are dealing with the similarities and differences among organisms. They use these patterns to identify kinds of organisms, and they also use them to develop branching models of the evolutionary relationships among those kinds. But in systematics, more than any other sector of biology, we find a severe clash between categories and an understanding of evolutionary processes.

As shown in previous chapters, evolutionary groups do give rise to the patterns of similarities and differences that we find, but those groups often correspond poorly to the species that we define. The poor match between evolutionary groups and recognized kinds is, of course, a terrible difficulty for anyone who would study the evolution of *kinds* of organisms. Defined species are not the same as the entities wherein evolution happens. As much as the taxa that we identify (including species taxa) are caused by evolutionary groups, they are also caused by our impressions of recurrence—they are categories first and foremost. Even in those cases where species taxa do correspond well to evolutionary groups, it is far from clear that the history of those groups can be estimated well simply from patterns of similarities and differences —the types of observations that were used to define the species. Evolutionary groups can have quite complicated histories, far more complicated than can be depicted with a simple branching model. Unfortunately the major goal of modern evolutionary systematics is to develop an accurate branching tree model of the evolutionary history of kinds of organisms.

The Divorce of Classification from Evolution

Of course, systematists are very familiar with how difficult it is to identify and measure species. Indeed a fair portion of the history of systematics has been a preoccupation with the species problem. When one does not know how to identify species or to define SPECIES, it is very hard to categorize them and to study their evolution. Similarly when different scientists cannot agree about species and SPECIES, then the results garnered by those who use one set of methods will be distrusted by others who don't agree with those protocols.

Given that over the years the species problem has often been seen as inherent and intractable, some have argued that systematics should simply abdicate the goal of obtaining accurate estimates of evolutionary history. Certainly, in a simplistic sense, much of the species problem would not arise if biologists stuck to devising categories, and to placing organisms in those categories. The argument is that rather than spend so much energy on false hopes of good, branching, supposedly *true* phylogenies, that our primary need is for a practical working method of classification and of organizing taxa. Under this view, a systematist should not focus on the impossible task of reconstructing evolutionary history, but should turn instead to applying a workable method that will lead to a useful summary of biological diversity. This is a seductive argument. The difficulty, and it is a terrible one, is that once we abdicate the goal of matching categories to reality, we no longer know by what criteria to judge whether a systematic method is useful (Baum and Donoghue 1995; Bachmann 1998).

One of the most famous and fully developed arguments along these lines, was a justification for phenetics, a school of systematic thought · that proposed mathematical methods for grouping organisms based on measurements of similarity (Sokal and Sneath 1963). The basic argument for a reliance on similarity begins with the acknowledgment that any method of identifying groups of organisms, even the initial stages of taking a sample, requires an assessment of the patterns of similarity among the organisms to be studied. Realizing this, and also realizing the very uncertain nature of real evolutionary groups, Sokal and Crovello argued that assessments of overall similarity should be the basis of a practical taxonomy. They also were clear on an important corollary, that taxa identified on the basis of similarity patterns should not be confused with real evolutionary entities (Sokal and Crovello 1970).

A very similar argument was adopted by some advocates of the cladistic method of fitting branching models to phylogenies, that was inspired by Hennig's book, *Phylogenetic Systematics* (Hennig 1966). The central idea follows from that of "descent with modification" (Darwin [1859] 1964, 123), that as a species evolves it acquires new traits, and that when it splits into two species, both sister species will carry those traits that were held by their single recent ancestral species. Thus, the branching pattern of species history could be reconstructed from the pattern of shared traits. The cladistic approach is elegant, and popular, but it has not dispelled the many uncertainties that arise when one tries to study the evolution of kinds of organisms. Among cladists, the species problem has flourished just as it has for all systematists. But some systematists, the so-called pattern cladists (Beatty 1982), conceived that the cladistic method of constructing branching relationships was ideal because it was neutral with respect to evolutionary biology, particularly with respect to the more problematic and uncertain aspects of evolutionary biology. The pattern cladists embraced the simplicity of the cladistic method, and claimed that it could best be used without reference to history, that relationships among groups of organisms could be constructed just as well without any reference to evolution (Platnick 1979). Thus, they undertook the same argument made earlier by the pheneticists, that in the face of great difficulty construing real evolutionary histories, and matching taxa to them, it is best to focus on classification and organizing taxa by methods that were not sensitive to the species problem and the difficulties of figuring out evolutionary history. They argued that what was special about cladistics, as opposed to other methods (particularly phenetics), is that it is the simplest, most assumption-free, method for organizing taxa. The philosophical justifications of cladistics were intended to overcome any skeptical concerns that, since true evolutionary history was no longer taken to be the organizing principal, that any and all systematic methods are equally valid.

Of course, the problem that arises when one divorces the process of classification from the study of evolution is that nobody is very interested in the resulting classification. Other biologists do very much want to know about the real evolutionary history of the species they study. Biologists want the names of kinds of organisms, that make up a classification, to correspond to real entities in nature. Whether or not their wishes are reasonable, they do not want to be given something else and be told that their wishes are not reasonable. Those who have emphasized the practical utility of simplistic methods of classification have overlooked the fact that other biologists are not really interested in classifications that deny the evolutionary processes that gave rise to diversity. One may argue the merits of practicality, but in fact, and as a practical matter, systematists have not been allowed to forget their knowledge of evolutionary processes.

In effect, systematists are damned either way. If they say that it is best to avoid claims about real evolutionary history, then others say they are ignoring what really matters. If they turn and embrace reality and try to connect taxa to real evolutionary processes and history, they are back in the seemingly hopeless species muddle.

Encysting the Species Problem

Today there seems to be a growing consensus among many systematists that Hennig's methodological philosophy, that gave rise to cladistics, is indeed the proper basis for systematics. But rather than try to deny evolution or evolutionary history, many now embrace cladistics as the best way to organize taxa *and* to reveal phylogenetic history. The new phylogenetic systematics (PS) is now widely embraced as a means toward understanding phylogenetic history, and also increasingly as a means for classifying and naming (Cantino et al. 1997; de Queiroz and Gauthier 1994; Hibbett and Donoghue 1998). In practice, the basic analytical protocol of PS is the cladistic one. But rather than claim, as might a pattern cladist, that evolutionary theory is ignored in the tree building procedure, current practitioners of PS closely resemble Hennig and the original cladists in embracing the evolutionary process as the cause of hierarchical patterns in nature, and thus of hierarchical classification schemes.

Phylogenetic systematics has not solved the conflict between the demands for good classifications and well-understood history. But there has been an innovation of sorts. PS practitioners have folded their uncertainty over species into an ostensibly reduced debate about the best definition of SPECIES. With the major difficulty seemingly encysted, systematists can proceed by claiming that species uncertainty is not at the center of the endeavor, but rather exists on the fringe, having been delimited to a pragmatic choice among carefully crafted alternative definitions. Take careful note of this method of resolution, for there are two ways to describe it: by the first, it is clever; and by the second, it is injurious to our understanding of evolution. It is clever for providing a way to work and carry out systematics in a way that preserves many traditions. By setting aside the species problem, phylogenetic systematists can feel free to describe patterns and classifications among taxa and to ascribe branching historical models to them. The harm is that those historical models and the categories they connect and all that is built on them may be wrong for having overlooked evolutionary processes.

In the following quotes, note how the purpose of an ideal species concept—the best definition of SPECIES—is to aid in revealing evolutionary history and is not the result of an understanding of evolutionary history. "If our goal is to resolve genealogical relationships as finely as possible . . . then we need to develop a species concept and definition consistent with this goal" (Donoghue 1985, 180). "The goal of a phylogenetic species concept is to reveal the smallest units that are analysable by cladistic methods and interpretable as the result of phylogenetic history" (Nixon and Wheeler 1990, 211).

For many practitioners of phylogenetic systematics, what a species concept is *not*, is an idea of what real species *are*. Rather it is an idea that helps *identify* the real participants in the phylogenetic process. The method begins with the strong assumption that the process of biological diversification, viewed over time, has a truly branching pattern, where a branch is taken to be some sort of temporal sequence of a group of organisms. Given this, a phylogenetic systematist seeks the best criteria for identifying contemporaneous groups of organisms that serve the process of revealing the historical branching pattern. The final element of this logic is an argument for the best species concept. The best definition of SPECIES is the one that proves to be the best guide for identifying those criteria that prove to be the best for revealing the real history. Does this seem backward? Somehow a species is to be identified without being described in any other terms other than those used for identifying it—and yet, these species are supposed to be real. The motto seems to be: "We shall devise that which works best, and that will be the true one."

There is really just one mistake in this argument. I don't think it is a mistake in reasoning, exactly, but rather an error in carrying mistaken assumptions too far. Those who follow this PS credo would assume that evolution is simple, in order that they need not deal with the ways that it is not. In fact, the method is essentially that of most science, and indeed it is the method of this book. One must make some assumptions, even questionable ones, about some things so that one can make progress on other things. Unfortunately some would presume, not just as a starting point but also as the quintessence of systematics, that the only possible causes of biodiversity are discrete branching events of species. It is a kind of reductionism, but a corrupt one, as it demands that evolution really does happen to distinct things and that it really does cause distinctly branching histories. It is a doctrine that denies the knowledge that neither species, nor their events of splitting, are always distinct. With a misleading credo tucked close to the breast, PS practitioners go forth to solve the species problem by finding just the right words for a definition of SPECIES that causes the fewest problems.

By relying so strongly on assumptions that are known to be false, phylogenetic systematists are lead to some awkward positions. One is to claim that there can be two different kinds of reality, distinguished on temporal grounds: there is both an historical reality and a contemporaneous one. Biologists do tend to fall into two groups, based on whether they have a "snapshot" viewpoint or an historical viewpoint of species (Endler 1989). A contemporaneous view is most likely to be useful to a population geneticist or ecologist focusing on ongoing population processes, such as competition and recombination. In contrast, systematists take a historical view and generally refer to species within the context of ancestor-descendant relationships. For example, Simpson (1961, 150) included the phrase "ancestral-descendant sequence of populations" within the evolutionary species concept. Cracraft defined a phylogenetic species as "an irreducible (basal) cluster of organisms, diagnosably distinct from other such clusters, and within which there is a parental pattern of ancestry and descent" (1989). But a tendency for different perspectives by different biologists cannot be confused with different kinds of reality. The historical, though past, is necessarily a temporal extension of the contemporaneous. If there exists a kind of species that can be defined by historical relationships, then these historical relationships must have arisen because of processes that occurred over time. In both of these examples, Simpson's and Cracraft's, the definition of a historical species concept supposes the existence of some kind of cohesive group of organisms that exists for each slice of time in the history of the historical species. In short, any suggestion that both views of reality, contemporaneous and historical, can be sustained as distinct and valid must suppose two different sorts of reality.

The motive for treating historical and contemporaneous views distinctly is of course, that as soon as one envisions them as the same, one must embrace all of the difficulties of indistinct boundaries and fractal hierarchies that are well known as part and parcel of the evolutionary process. When one does so, history is revealed as not entirely branching, and classification is revealed as partly arbitrary.

Systematists do sometimes try to bridge the gap between historical and "snapshot" views, by considering the process that creates a contemporaneous species, and then extending that process through time to generate a picture of a historical species. Indeed much of the literature that forms the debate on the phylogenetic species concepts includes this very exercise (Baum and Shaw 1995; Davis and Nixon 1992; de Queiroz and Donoghue 1988; Hennig 1966; Nixon and Wheeler 1990). What happens then is that the descriptions of the contemporaneous versions often incorporate interbreeding, and thus end up sounding a great deal like the biological species concept (Claridge et al. 1997a) or they imply something that "contributes to a parental pattern of ancestry and descent" (Cracraft 1989, 35) and thus sound like Templeton's cohesion concept, which is largely a contemporaneous description (1989, 1994).

Much of the misunderstanding over historical views of biological diversity has been caused by a mistake in which an ideal of a four-dimensional view of life's happenings is confused with the pattern that has been left by history. We can only see history in brief glimpses, constrained by patience and memory, but we do try to picture it and to model it, as Simpson was doing when he defined species. What we can perceive, quite contemporaneously, are the patterns of similarities and differences that exist among current organisms. Furthermore, we appreciate that evolutionary processes caused those patterns. However the pattern is not the history, but merely evidence of it.

I do think that there is a useful distinction to make, one that is closely related to that of historical versus contemporaneous reality. Later (chapter 11), the case is made that many of the patterns that have been left by history, and that are used by systematists to devise taxa, are real (or at least, because they are so useful, they should be treated as real). They are not the same kinds of entities as evolutionary groups,

but both the patterns and the evolutionary groups are perfectly contemporaneous.

Another awkward claim that often arises in the context of phylogenetic systematists, is of a hard epistemic distinction between evolutionary pattern and evolutionary process. Once again, systematists have
drawn on a common practical distinction (e.g., this book often refers to
processes and to patterns), and taken it to be a rigid partition in the way
we come to understand the world. The claim is that one can study the
phylogenetic history—the supposedly true branching history that is assumed to have happened and to have lead to currently recognized
taxa—by examining present-day patterns and without delving into
the complexities of evolutionary processes. For example, Vrana and
Wheeler (1992) were rightly concerned that any attempt to describe
taxa that are placed at the tips of the branches of an evolutionary tree
necessarily invoked a great load of assumptions regarding evolutionary
process. Wishing to avoid this mess, they proposed that processes (and
taxa as well) should be explicitly ignored, and that the tips of estimated
phylogenetic trees should be individual organisms, with no reference
made to real evolutionary entities. Their justification was the same as
that of pheneticists and pattern cladists, who subjugated evolutionary
processes in favor of classification. Similarly, Chandler and Gromko
(1989) argued that since species arise by a variety of mechanisms, each
of which may lead to different contemporaneous ideas of species, that
a species concept should be distinct from ideas about what gives rise to
species. They say that what qualifies entities as species is independent of
what caused them to be species. The proposal strongly begs the questions of just what a species concept is a concept of, and to whom a
species may be qualified. It is a denial that we should use our knowledge of the evolutionary process to describe the results of that process;
that we should, instead, describe the results of that process as if it is
something else. When they wear such blinders, there is little room for
systematists to perceive and to incorporate into their understanding of
phylogeny the many circumstances where biological diversity is not
sharply partitioned, where real species are vague because of indistinct
boundaries or because they are nested as parts of larger species.

To better see how illusory are these tenets of phylogenetic systematics, consider what might happen if the field somehow prospered to
the point that its tenets were deemed so valuable as to merit placement
early in the course of a biologist's education. Envision a biology curriculum built upon such a foundation, and consider the student who
asks, "How do phylogenetic species come into being?" A practicing

phylogenetic systematist might well respond with something like "They come into being when they evolve diagnosable traits." Consider then, two fairly unpleasant questions any inquisitive student might follow with. The first would probably be "How do those traits come into being?" Could this question possibly be answered without releasing a flood of disquieting knowledge about the complexities of evolution, knowledge that the central presumptions of phylogenetic systematics are often false? The student, cynical by now, might well follow with a second unpleasant question, something like "If there is a species of tree in the forest, and no systematist is around to diagnose it, is it still a species?"

Phylogenetic Systematics in Practice

Consider a recent statement on the goals, purpose, and state of understanding of systematics, edited by Claridge, Dawah, and Wilson (1997b). This book contains many thoughtful chapters on a variety of diverse and traditionally "difficult" groups of organisms, and it has been well written and edited with an eye toward a calm, well-reasoned tone. In the opening chapter, the editors write of the maturity and of the auspicious future of phylogenetic systematics. But no amount of reasonableness can make up for the inherent conflicts within the goals of the field. Throughout, the book is thoroughly penetrated by the conflicting wishes for improved understandings of biological diversity, and for unambiguous categorical representations of that diversity. (Interestingly, it also contains an article by David Hull that describes the different, and inherently incommensurable, goals of systematics (Hull 1997)). In some cases, we plainly see the conventional dualism, where authors envision species as both real entities in nature, and as categories that we construct, and yet fail to see the inherent conflict. Thus Van Regenmortel writing about viruses says that "A virus species is a polythetic class of viruses that constitutes a replicating lineage and occupies a particular ecological niche" (1997, 18). A polythetic class is just a category in which the members need only meet some, not all, of the membership criteria. Though more flexible than a classical rule-based category, it is still a constructed abstraction, and yet it has been equated with supposed distinct entities in nature. This finely crafted statement simply assumes that two quite different things—one is some portion of organismal diversity that occurs outside of ourselves, and the other is a constructed category that exists in the mind—can be equated. Several other chapters appear at first to avoid this confusion by focusing strongly on the practical. This was the case for the chapters on lichens (Purvis 1997), eukaryotic algae (John and Maggs 1997), nematodes (Hunt 1997), and cultivable bacteria (Goodfellow et al. 1997). All of the authors maintained that real entities in nature are hard to discern or are truly indistinct, and that constructed practical ideas behind the use of SPECIES must continue to be used. But there is a strong existential thread running through each of these chapters, in the way they refer to species. They each make reference to the number of real species in nature. Thus for cultivable bacteria we learn of the practicalities: "species level taxonomy is based explicitly or implicitly on detection of morphological discontinuities in sets of field-collected or cultured algae" (John and Maggs 1997, 84). And we also learn of the realities: "The number of species presently recognized . . . is estimated to be about 10% of the true worldwide total" (John and Maggs 1997, 84). The article on lichens, which throughout emphasizes the practical difficulties of identifying lichens, also asserts that "a conservative estimate of the total number of species would seem to be 30,000" (Purvis 1997, 111). In another chapter, we are told that "the Phylum Nematoda comprises a diverse assemblage with tens of thousands of nominal species. Estimates of the total number of species are speculative, but range into the hundreds of thousands" (Hunt 1997, 222). Finally, in the chapter on culturable bacteria, we again see both sides: "recent developments in bacterial systematics are being used to provide an improved operational species concept . . . However, it must be remembered that the number of bacterial species known and described represents only a tiny fraction of the estimated species diversity" (Goodfellow et al. 1997, 27).

Each chapter asserts that practical meanings of SPECIES must serve at least for the time being, and yet each reports an estimated count of the number of supposedly real species in the world. In each case, the authors are trying to answer the call of two imperatives: the demand for names and workable classifications for the organisms they study and the demand for understanding of diversity in the real world. Being good researchers of their organisms, they recognize that they cannot simply equate the categories that they construct with real world entities, yet they remain compelled to do exactly that.

Of course, there is a way to reconcile the apparent conflict that lies within these statements. If in each case what the authors were referring to in nature were not supposed true entities, but rather what would be counted by systematists if they persisted in constructing categories for all the remainder of undiscovered organisms, then their statements would be consistent. That would be a defensible position in some respects, but it would also entail a much stronger statement on the divorce of species in systematics from entities in nature. It would strongly affirm the common notion that so many biologists assert as repugnant, that *a species is just what a systematist says it is*. None of these chapters explain their worldwide counts in any such terms, and each of these chapters profess a wish that in the future, the current practical ideas of species should move to correspond with (hopefully improved) understanding of entities in nature.

In the opening chapter of the book the editors write of a growing and improving tradition of phylogenetic systematics that relies on cladistic methods of estimating phylogenetic history, and that draws upon the phylogenetic species concept (Claridge et al. 1997a). They write of the compatibility between the biological species concept (BSC) and the phylogenetic species concept (PSC) in many contexts, but they also emphasize the diversity of mechanisms that can give rise to real species in nature and that are not all recognized by the BSC. They also embrace a strong distinction between knowledge of pattern and knowledge of process. As discussed above, the distinction is illusory once one has some insight to both. The pattern of phylogenetic history was shaped by processes, and we have a great deal of evidence that some of these processes contribute to patterns of branching and that others contribute to more complicated kinds of historical networks. None of the kinds of data that are used, nor any of the methods for estimation or analysis or data gathering, can be selected or evaluated without drawing on what knowledge exists of evolutionary processes. To try to do so means that one necessarily embraces a simplistic view of evolutionary processes, one that is inconsistent with that state of knowledge. For good reason, most scientists will repudiate such dumbing down. They may be motivated in conflicting ways when it comes to species, but most are also motivated to use as much knowledge as they can draw on in order to gain more knowledge.

In support of their contention that pattern can be studied, while ignoring process, the authors write: "Our understanding of patterns of diversity in the field should not vary according to which theory of speciation we support." Note how the statement shifts from the seemingly sensible to the irrational, if we replace the last two words, "we support," with "we find evidence for."

A central theme of the volume is the utility of a version of the PSC as proposed by Cracraft (1997). Actually, this particular version belongs to but one of two strongly competing schools of thought on what should be the ideal PSC (Wheeler and Meier 2000). One school holds that, among other things, species are monophyletic groups (Mishler and Theriot 2000), whereas the other does not require monophyly (Wheeler and Platnick 2000). Cracraft's version falls in the latter category (1997), and for present purposes it will suffice (see chapter 10 for a discussion of monophyly in the context of species concepts).

Before presenting his version of the PSC, Cracraft writes with striking and commendable clarity on the purpose of an idea about species, and the problems for systematics if ideas about species are not considered to correspond to real aspects of nature. "Unless species concepts are used to individuate real, discrete entities in nature, they will have little or no relevance for advancing our understanding of the structure and function of biological phenomenon involving those things we call species " (1997, 327). "If species are not considered to be discrete real entities . . . then it implies that evolutionary and systematic biology would be based largely on units that are fictitious, whose boundaries, if drawn, are done so arbitrarily" (1997, 327).

The actual definition of a phylogenetic species is: "the smallest population or group of populations within which there is a parental pattern of ancestry and descent and which is diagnosable by unique combinations of character states" (Cracraft 1997, 329).

Readers should see the problem immediately. The statement equates the categories that are discerned, or diagnosed, with real individual entities. It is blind to the possibility that the diagnosis may yield something that has no direct, distinct real counterpart in the world. As a refined capsule of the species problem, we could hardly do better.

Cracraft's latest version is just one in a series of attempts by he and other phylogenetic systematists to craft a series of words so as to say two incompatible things at once. The statement is perfectly suited to appeal to those who feel that species are real things in nature, and that they are things we must be able to distinguish. Sadly, there is simply no room in a diagnostic protocol, based on characters that appear distinct to human observers, for real aspects of nature that have truly partial boundaries, and that occur in truly fractal hierarchies. The problem is much worse than square pegs and round holes, rather it is like trying to put clouds into boxes, and still have them be just as they were.

In addition to the built in epistemic conflicts, the PSC also suffers badly in other ways. A very strong charge against it is that it could be used to diagnose even the smallest groups of organisms, or even all individual organisms, simply because organisms have an effectively infinite number of characters (Avise and Ball 1990; Mallet 1995). Indeed, this very difficulty has emerged in practice (Packer and Taylor 1997; Peterson and Navarro-Siguenza 1999). Cracraft's answer is that "the PSC is not about the diagnosability of individuals but of populations" (1997, 330). This response, like the PSC itself, necessarily entails transferring all of the semantic and empirical confusion of SPECIES onto POPULATIONS: what is a population?; how do we know where its boundaries are?; and how do we first identify it so as to check its diagnosability? (The same problem arises under the Biological Species Concept— see chapter 12). Furthermore, the PSC still overlooks the vast amount of variation that occurs among organisms and their genomes (Avise and Ball 1990). If what matters is diagnosability, and if characters can be found virtually without end, then what is to prevent the finding of characters that help one to diagnose any collection of organisms?

The Concept Problem

It is a considerable convenience, for those concerned with how we refer to biological diversity, that much of the species problem discussion that is engaged by systematists has been crystallized into debates over how to frame a species concept. For now, it is clear that our very wish to design the perfect concept lies close to the heart of our species difficulties. The idea that we shall construct a meaning, where nature will not serve, might seem reasonable given the difficulties of interpreting nature. But we see now, in retrospect, that the striving for just the right, short definition has been caused by a very strong wish to have our categories match up with real evolutionary groups, and a failure to appreciate that that cannot happen. Those who would define SPECIES in a way that somehow finesses the mismatch have undertaken an impossible task.

Recall the point made in the first chapter, that we biologists who suffer and debate the species problem have not been asking for more data or for more knowledge about biological diversity. We think that we are full up with the knowledge of biology, and that what remains is to shape it somehow so as to make the species problem go away. We have been mistaken. What we have missed is an appreciation of our own role in devising categories, and of our own desires to have those categories be the entities in our theories. Evolutionary groups are just one major cause of our species taxa, and we are the other.

This chapter has focused on systematics, a discipline long known for suffering the species problem, and a field of not a little infamy for hosting some dogmatically polarized debates (Hull 1988). But systematists are far from being the only biologists who suffer the species problem, as we shall see in the next chapter. In fact, a good case can be made that much of the disputatious discourse that happens in systematics is a result of the impossible species-related demands made by other biologists. Virtually all biologists, at some time or another, come to a need for an understanding of the evolutionary history of the organisms they study. And not only biologists, but also a great many other people, are steadily in need for identification of organisms. In various ways, society demands that biologists have at the ready, names and evolutionary histories for all the organisms on the planet.

TWELVE

EVOLUTIONARY BIOLOGY

Apart from systematics, there are other areas of the life sciences where practitioners regularly suffer the species problem. These areas are primarily those in which the research for one reason or another focuses on organisms as we find them in nature, and on the evolutionary processes that have caused them. There is no very good single name for all these efforts, though sometimes they are collected under EVOLUTIONARY BIOLOGY (a label I will use hereafter), or POPULATION BIOLOGY. Certainly much of what is referred to with ECOLOGY falls in this domain. Systematics could be also be included, but that field tends to play out as a fairly distinct subdiscipline of its own, and here I am making a distinction between systematics and an area of biology where the research focuses more explicitly on ongoing evolutionary processes. The distinction is a common one, and useful for the present purpose, but as we shall see, in the end all evolutionary biologists, systematists included, suffer the species problem in a similar way.

The debates of species concepts tend not to be so sharp in evolutionary biology as in systematics. Part of this difference is illusory, as I have simply defined EVOLUTIONARY BIOLOGY as pertaining to a broader field than my usage of SYSTEMATICS. But it is also within evolutionary biology where the mechanisms of evolution, as they play out in the short term, are studied, and thus there is a strong tradition of recognition among evolutionary biologists that evolution does not always give rise to distinct groups (Howard and Berlocher 1998). Thus, evolutionary biologists are not so likely as systematists to be confused by the lack of distinction that is often found in nature. But evolutionary biologists are motivated to identify and employ names for organisms, just as are systematists, and collectively they are probably the greatest source of demand for ready names and phylogenies of organisms. As such they are to blame for much of the pressure on systematists for the rapid placement of organisms within kinds and for the placement of those kinds within hierarchies.

Biologists in the Field

The way that typological thinking and presumptions of distinction play out for evolutionary biologists are best appreciated by considering the basic endeavor of going out into nature and collecting a sample. Suppose that species exist as real evolutionary groups in nature, and that one would like to collect some organisms to see if they are really a part of such a real species. Immediately one is faced with a considerable and inescapable difficulty, though it is generally not recognized. It is the difficulty of answering this question: "How can we sample organisms, for the purpose of studying a species of which they are a part, without first knowing what they all look like and where they all are?" If a species is an individual, an entity in nature, then each and every organism in the species is a component of that species. An investigator cannot knowingly collect every organism in a species, because she does not yet know which organisms are part of that species. She can work with a sample of organisms, but a sample must be drawn from a widespread area, including the full range of the species to be identified. But until she identifies the species, she does not know how widespread it is, and she cannot identify it until she has a widespread sample. You might say that she could find evidence of a species with even a small sample, and this is true; but she would not then know where the species occurred. This is a nasty problem without a logical solution short of sampling everything.

Let us take the proverb of the blind person and the elephant one step farther. The blind person could identify the elephant if he crawled all over it *and* if he had some prior notion that the elephant had boundaries where it joined the ground. But lacking any prior notion of the form or location of boundaries, he would not know when his search for the elephant had ended. Evidence that this conundrum plagues biologists is easily found. The history of species studies is littered with cases where a single species, described and named on the basis of a small local sample, is later found to include multiple species. In short, it is extremely difficult to identify individuals (any kind of individual) if you can see only its parts. A priori, one simply cannot know how many parts it has and where those parts occur.

The difficulty of trying to describe real species, by sampling from them, is one cause of creeping species counts. Recall from chapter 2 how a fundamental and recurrent type of failure of species measurements is that the larger our sample, the more new species we find. The problem is not the association between sample size and numbers of observed species, rather it is our tendency to report counts of the categories, devised on the basis of small samples, as reasonable estimates of diversity. A second cause of creeping species counts is that real species can be expected to be indistinct and to have fractal hierarchies of species within species, and so are not countable. Such is the case with the parts of fractals generally (chapter 6). Thus even if one reports counts of only the number of evolutionary groups for which evidence was found in a sample, that number could be expected to grow, perhaps indefinitely, with sampling effort.

In practice, when biologists sample organisms and consider whether they are a part of one or another species, they do not do so by considering the individualistic nature of species. They do not first inquire of the forces that cause evolutionary groups. They do not ask, "Have these organisms been engaged with others such that they all form an evolutionary group?" There are important exceptions, and evolutionary biologists are very aware of the distinction that is made here—it is the difference between assessing an organism's actions and effects within its environment, and assessing what it looks like—but the fact is that the vast majority of decisions in which an organism is identified as belonging to a species, are decisions in which species are just categories, identified on the basis of similarities just as for any natural kinds. The typical protocol is induction: first, to sample organisms, then to draw conclusions from that sample, and finally, to extend those conclusions to other organisms that were not in the sample. In practice, inductions work much better when one has a well-described and well-constructed natural kind (this is true of any natural kind). So in the context of large samples that have been used to define a natural kind, inductions made from those samples will often be born out, more often than when the natural kind is based on a small sample or a poorly studied sample. The problem is that if species are individuals, then they are not natural kinds, and inductions may fail for all sorts of reasons that have nothing to do with the individuality status of the samples that inspire the induction, or the individuality status of the larger group to which the induction is applied. In short, there is no logical way of characterizing a group sufficiently such that its individuality can be unfailingly assessed with a sample. Without a presumption about the limits of an individual, one cannot ever know that it has been entirely examined. The best one can do is to assess the individuality status of one's sample. So it must be for real species as well. Regardless of how clearly a sample reveals evidence of being part of an entity, application of individuality status to anything beyond the sample is necessarily via induction. Furthermore, induction works well on natural kinds but poorly on parts of individuals. This general difficulty has troubled and will continue to trouble biologists to no end, so we may call it the species-sampling problem.

The species-sampling problem can be overcome in one way, and that is to restrict all conclusions on the evolutionary status of a sample, to just that sample. This is what is done when a sample comes from a relatively unexplored context where it is not known where there occur, or even if there exist, other similar organisms. Modern examples include the collections of microscopic organisms from sediments, or extreme environments, where the investigating biologists bring relatively little preconceptions about what they may find (Embley and Stackebrandt 1997; Munson et al. 1997; Vetriani et al. 1998). But the investigators of difficult and novel microbes have their logical formalism imposed upon them. Most biologists are working in contexts where others have gone before and have devised categories on the basis of their samples. When another investigator comes along, new samples are taken with the aid of a series of search images, such as field guides, or a series of criteria, such as identification keys. In the case of search images, organisms are identified to kind in a typological way, and in the case of distinct criteria, organisms are identified to kind as if kinds are classical categories. When an organism is found that fits neither search images, nor catalogs of criteria, then the typical reaction is to consider it as a representative of an unknown species. All of these behaviors treat species as categories, and not as entities in nature. They are practical time-honored methods, but they necessarily leave biologists inadequately disposed for handling situations where their categorical kinds of organisms are not good approximations for the real evolutionary landscape they are studying. The best that evolutionary biologists can do, to guide them when they must refer to organisms that are not clearly part, or are clearly not part, of distinct evolutionary groups, is to restructure their categories to better fit the real evolutionary entities they find. Another common practice is to use a taxonomic rank even lower than species, such as subspecies. There exist no theories on the nature or origin of subspecies. Rather it is simply a designation that seems necessary when a distinct kind of recurrence is found among a subset of organisms otherwise recognized as one species (Winston 1999, chap. 17).

Evolutionary Processes and Species Concepts

Evolutionary biologists, like systematists, have also been building a dictionary for SPECIES. From Mayden's listing (1997) we find at least ten species concepts (SCs) that either directly invoke evolutionary processes or clearly imply a role for evolution in causing or maintaining species, including the Biological SC, the Cohesion SC, the Ecological SC, the Evolutionary SC, the Evolutionary Significant Unit concept, the Genetic SC, the Hennigian SC, one version of the Phylogenetic SC, the Recognition SC, and the Reproductive Competition SC. All of these concepts say something about what holds a species together, and thus they are all directed at real entities. At least in their written summaries, these concepts exhibit little confusion between what exists in nature, and the categories we devise. It is for this very reason that they are not popular among systematists. These concepts are seen as not being "operational" because they do not provide strict criteria about when and where there is a species. But evolutionary biologists don't want just a theoretical notion of species, they also want something that helps them to identify real species. In fact, evolutionary biologists would place even greater demands on a concept than would systematists. They want the concept that explains real species, *and* they want that explanation to provide an operational criterion for identifying species. In contrast, phylogenetic systematists are inclined to say that whatever turns up, while the investigator is using the best operational concept, are also real.

Without doubt, the species concept that has received the most attention by evolutionary biologists is the Biological Species Concept (BSC, see chapter 7). At least on the surface, or as a statement in plain text, it describes what goes on within species and what does not go on between them (sexual reproduction in both cases), and at first glance it also describes how to decide which organisms are in the same species (i.e., assess which organisms engages in sexual reproduction). As described in chapter 7, these are important ideas that have been around a long time, well before they were laid out explicitly as a species concept by Mayr. They also correspond fairly well to some expectations that came up in the discussion of sexually reproducing, or otherwise recombinogenic, evolutionary groups. But many biologists—systematists certainly, but also many evolutionary biologists—are entirely unsatisfied with the operationality of the BSC. A major concern is that the BSC does not serve for organisms that never or rarely engage in sexual reproduction or exchange genes with other organisms. But even if we ignore that major shortcoming, there remains an enormous practical difficulty. Most times when biologists would examine organisms in order to make decisions about species, it is practically impossibility to make assessments of whether or not they do, or can, share in sexual reproduction or recombination.

Many authors have commented on the practical difficulty of the BSC as a guide in identifying or classifying organisms or as an aid in figuring out the evolutionary relationships among organisms. Mayr (1996) has two responses to these criticisms of nonoperationality. First is a repudiation of those who would impose a categorical view on the real entities in nature. He asserts that the BSC is simply a description of the processes associated with real evolutionary entities and that these may be indistinct at times. At such times, he recognizes that understanding species is a better goal than is counting or identification, and he seems to deny any imperative to draw or impose distinctions where none exist to be found. So far so good. It is exactly this sentiment that I have learned, partly from Mayr, and used throughout this book with regard to evolutionary groups. But Mayr's second rejoinder to systematist critics is that partial boundaries are generally rare, and that most of the time one can identify and count species without difficulty. These two rebuttal points are awkwardly juxtaposed by the common question of how to consider allopatric populations of very similar organisms. Earlier versions of the BSC included "potentially interbreeding" but physically separated populations, as part of the same species (Mayr 1942). In later versions, POTENTIALLY was dropped and instead Mayr says that with allopatric populations one must infer, on the basis of nonreproductive features, whether to count isolated populations as one or two species (Mayr 1992, 1996). Thus, on the one hand, Mayr writes as a population biologist focused on evolutionary processes when he asserts that species are entities by virtue of reproductive processes. Yet he is also certain that some allopatric populations that do not share in reproduction are of the same species: "when one compares allopatric populations one is apt to find a mixture of strictly conspecific populations, incipient species, and good species" (Mayr 1992, 224). In other words, two entirely separate populations that have zero gene exchange, two completely separate evolutionary groups, are sometimes both part of a single real species. Why do allopatric populations of similar organisms lead Mayr to abdicate the process of gene exchange, to abandon his theory in favor of demarcation? The reason is that he would like the BSC to be useful, not only to evolutionary biologists studying single populations, but also to those who would count and classify species.

Mayr wishes that the BSC would serve two purposes: as a description of what real species are in nature and as a tool to serve for identifying species. Though they swap these two priorities and say that what is identified is real, phylogenetic systematists want the same thing. Neither the proponents of the BSC, nor the PSC, are peculiar. In the actions of both we see what most biologists desire: We want to be able to identify distinct categories of organisms, and we want those categories to match up with real evolving species. This wish gives us our species problem.

In practice, the BSC rarely serves for identifying species. This is awkwardly but well revealed by Mayr's attempt to show the utility of the BSC for plants. In response to claims that plants were especially problematic for the BSC, that they offered many cases when reproductive isolation was not a good guide for identifying species, he undertook to examine the vascular flora found near Concord, New Hampshire, to see how many species the BSC did indeed fit. Of 837 species, Mayr concluded that the BSC fit 760 of them quite well (93.6%). However, every single assessment was based entirely on a finding of distinct morphology. No attempts were made to assess hybridization or gene flow, even though many of the species were known to produce fertile hybrids with one another. Far from supporting the utility of the BSC, Mayr avoided it (Templeton 1998; Whittemore 1993). He did so for exactly the same reasons that most botanists do: The BSC does not help in consideration of the seemingly separate distinct kinds that do exchange genes, and it is impractical. Mayr's defense is that he was not avoiding the BSC, but rather just falling back on to a morphological criteria of species because it serves as a useful indicator of reproductive isolation (Mayr 1996). This too is just what many biologists do: They hold a theoretical idea of species in their head and then proceed to use similarity as a working criteria. Ironically, Mayr's botanical excursion served perfectly to make the point that the BSC is a theoretical idea and not an operational one.

The BSC is indeed a wonderful theoretical idea, one that does motivate a large amount of productive research. The absence of gene exchange among many evolutionary groups really does allow them to diverge from one another—so says the theory, and it is certainly consistent with all the commonplace and experimental evidence that similar organisms often can engage in sexual reproduction, while dissimilar organisms cannot. Thus, the evolution of barriers to gene flow is generally considered to be an important force shaping biological diversity. Evolutionary biologists know this, and that is why the evolution of reproductive isolation has been the subject of a great deal of research, since at least the early work of Dobzhansky (1936). But reproductive isolation is just part of this picture. The short BSC, as a theory, offers little for our thinking in those contexts where gene exchange is not even a possibility (under allopatry, or when organisms are not recombinogenic). Nor does it help very much in those situations where organisms turn out to be entirely capable of reproduction with one another, but for some reason choose not to. It also does not, by itself, explain why organisms that do share in reproduction also share in so many other things (though a line of reasoning like that in chapter 7 will serve). Finally, reproductive isolation is far from being an all-or-none feature, and a great many closely related species (by other criteria) are capable or have shown evidence of producing fertile hybrids. Evolutionary biologists know all about these things, and they have theories regarding them. They might even be considered to be shortcomings of the BSC, which is after all just a single sentence, but they are not problems for a broader theory of evolution.

Perhaps the greatest uncertainty of the BSC, that it shares with almost all other concepts with claims to being operational, is its reliance on POPULATION. The vagueness of this word is both a theoretical and practical shortcoming, and it is one that is generally overlooked. Mayr sidesteps the uncertainties entailed by this word, though he recognizes that the word must generally be vague (Mayr 1987). Even when asserting that "the task of the biologist is to assign these populations to species" (Mayr 1996, 266), he does not indicate what a population is, or how one should assign an organism to a population, or identify a population. Most other species concepts also rely upon POPULATION or something like it, which may explain why many critics of the BSC have not focused on this shortcoming. The vagueness of POPULATION may be necessary as is the case for so many words, but then that vagueness necessarily pervades any species concepts that relies on POPULATION. In general, a vague word need not be a difficulty. Whether part of a flexible concept or if used in reference to indistinct aspects of reality, a vague word may enhance communication by its vagueness. But if a definition is promoted as an unambiguous tool for identification and classification—for species or anything else—then it cannot be made up of vague words.

Over the years, the BSC has been revised several times, just as the PSC has been revised several times. Behind both endeavors lies a wish for as little ambiguity as possible and a basic misunderstanding of the place of definitions within knowledge. Knowledge is not contained in a dictionary, but rather in complicated networks of thoughts; nor is it conveyed with definitions—at least not much of it—but rather with full explanations of complexities and relationships. As much as biologists strive for unambiguous language, so that they may be understood (Keller and Lloyd 1992), a great many of their terms are necessarily vague, purely and simply because their referents are indistinct.

The Species Problem—Reprise

In the details, the species problem plays out a little bit differently for systematists and evolutionary biologists. But at base, both sorts of biologists are caught by the same conflicting motivations. People are motivated to refer to, and to understand, biological diversity. Some of this motivation may be part of the modern scientific tradition. But we know from studies of traditional cultures that people around the world have endeavored to develop taxonomic systems that permit them to quickly give names to the organisms they encounter, and we know too that these taxonomic systems are hierarchical on the basis of similarity. Today we see these two motives pitted against one another in a conflict that has resulted, ironically, from research on why organisms appear to fall into hierarchies of kinds. We continue to insist on names for kinds of organisms, and we continue to insist that those names fall into a hierarchy that fits the patterns we find in nature. But our best efforts at understanding nature have shown that real evolutionary forces give rise to entities that often will not fit ideas on distinct kinds of organisms. Our research has revealed the causes of biological diversity, but it has also revealed that our categorical tendencies are out of sync with those causes.

The organisms we recognize as belonging to the same species appear to us first as recurrent aspects of nature. And because of the way the mind recognizes and interprets recurrence, we must construct categories and refer to organisms as members of categories. We do so even when our insight to evolution leads us to understand that, though organisms often constitute real evolutionary groups, that these groups are often indistinct and may not match our categories. Finally, we are caught in a dilemma about how to refer to the recurrence that has inspired a category, yet was caused by an entity that is hard to discern. The dilemma has become crystallized within the struggle to craft a definition of SPECIES that would serve both our need to delimit the recurrence that caused a discrete category and our need to recognize the real entities that caused but may not match the category. Systematists and evolutionary biologists have both placed incommensurable demands on a word—that it do two very different things simultaneously.

THIRTEEN

WHAT ARE SPECIES? WHAT ARE TAXA?

The first of the two questions in the title of this chapter was set aside back in chapter 6 simply to avoid whatever existential presumptions might arise in trying to find an answer. As we return to it, we should keep in mind two things. The first is that, whatever else species are, they appear in our language as categories of organisms. The second is that, apart from the pleasantry of a short phrase, there is no need for a terse definition of the word SPECIES. Whatever real species are we can take our time to describe the reasons they exist and those things that different species share. Biologists' demand for a definition that cuts cleanly through species uncertainty may be understandable given the awkwardness of all that uncertainty; but as a hopeful mechanism for ridding ourselves of the species problem, it is misplaced. Quite simply, we cannot presume to describe things as having more distinction than actually occurs in nature.

Chapters 6 and 7 described the theory of evolutionary groups. Such groups of organisms are entities in nature, though we can expect them to be indistinct and nested within one another. Is the idea of an evolutionary group what people have been looking for in their search for a description of real species in nature? It is close in some ways. The general idea of an entity within which the forces of evolution occurs is one that shows up in many species concepts. But return to the example of dogs that began the first chapter. Their familiarity permits them to be a good example of just how jarring it might be to simply equate EVOLUTIONARY GROUP and SPECIES. Most dogs that are human pets or commensals are in one large evolutionary group, for dogs do move around the globe with people and compete with one another for people and other resources, and there is a fair bit of gene exchange across the group because of sexual reproduction. However, the larger group is very diffuse and it must surely whither into nothingness between many isolated populations of dogs. The group may also be blurred because of a low rate of gene exchange with wolves, which compete with humans rather than for access to them (Vila et al. 1997, 1999). Many dogs are also part of smaller evolutionary groups that we call breeds, and within breeds there are regional sub-breeds and these are also evolutionary groups. So too are there evolutionary group of dogs in Barrow, Alaska and in Madagascar, to pick two isolated geographic locals within which (I suppose) dogs exchange genes more so than with dogs that live else-where. Finally, we must suppose that Laika, the dog that was rocketed into space aboard Sputnik II in 1957, was in not in any evolutionary group while she was alive up there.

What we see here is a strong focus on the locals of evolutionary processes. To the concern that perhaps I have made total hash out of the notion of a group, there are two responses. First, we are considering a particular kind of group (i.e., GROUP appears with an important adjective). One of the most common ways that adjectives are used is to remove the sense of distinction that is caused by a noun alone, and so it is in this case with the adjective EVOLUTIONARY. Second, I have no choice. Evolutionary processes do play out among multiple organisms, and they do not do so uniformly across all organisms on the planet. How else are we to refer to those locals, and the organisms that consti-tute them, but with a word. And yet having used a word, it might seem that we also intend to convey distinction.

Recall the example of CLOUD in chapter 2. This is a word that is widely used to refer to things in nature, and yet those things are also universally recognized as sometimes being distinct and sometimes being indistinct. The cloud category is a conventional natural kind, one for which we recognize things as members by their similarity and by their recurrence in nature. We have devised a word for that category, even as we recognize that only some instances of that kind are distinct. We should also recognize evolutionary groups as a natural kind (i.e., the category of evolutionary groups, not any one evolutionary group). Like an individual cloud, an individual evolutionary group is itself not a kind or a category, but an entity. Just because they are indistinct or nested within one another does not mean we shall not refer to them as instances of a category. So many of the common categories we draw on for describing nature have some or many members that are indis-tinct in various ways: diseases, storms, solar systems, emotions, atoms, sneezes, and theories, to name a few.

Returning to SPECIES, we are left with a dilemma. On the one hand, we cannot have a single meaning that entails both recognized categories and real entities in nature. If we take one obvious step, and try to divorce from SPECIES any categorical meaning, and leave the word strictly for real things in nature, then what might those things be? I have just proposed that our best understanding of real entities in nature has been summed up with EVOLUTIONARY GROUPS. But I think it is premature to suggest this be used as a synonym for SPECIES. One reason is that to equate the two terms might entail a shift in usage—not for SPECIES, which is entrenched, but for EVOLUTIONARY GROUP. We have a theory for evolutionary groups, and the utility of that theory would suffer if EVOLUTIONARY GROUP ended up being equated with meanings of SPECIES that it does not fit. A shift in the other direction, in which people came to use SPECIES to refer just to real evolutionary groups would address the issue of what to do with the word. But I suspect that most people would find it difficult, or unsavory, because of the multiplicity and minuteness of many evolutionary groups.

There is one route that would preserve SPECIES for real entities and that would not equate it with all evolutionary groups, and that is simply to reserve it for large evolutionary groups. That is what I think we should do. Of course, such entities will also have indistinct boundaries, and the adjective LARGE does not delimit a distinct subset of the evolutionary group category. And it should go without saying that we could never entirely presuppose just how large a group might be, or that there could be any necessary correspondence between recognized species, in a categorical sense, and a particular size range or degree of distinction of evolutionary groups. In response to the concern that the idea of big evolutionary groups seems too vague, it must be said again that we cannot presume more distinction than actually occurs. If we are to have a word for recurrent, real aspects of nature, then we must admit that the word will be at least as vague as those real aspects of nature. So it is with evolutionary groups, even the large ones. There is no harm, nor paradox, in embracing a term that conveys an idea of potentially uncertain boundaries when in fact that is precisely the nature of the things to which we refer.

Monism and Pluralism

Biologists and especially philosophers of the species problem, sometimes wonder whether it is best to consider that there should be just one kind of species, as implied by the single word SPECIES (monism), or whether it is better to acknowledge the diversity of species concepts, or the diversity of modes of speciation, by acknowledging different kinds of species (pluralism) (Dupré 1999; Hull 1999; Mishler and Donoghue 1982). Monism is the default position for most biologists. We have one word that we puzzle over, and most of the discussion around the properties of real species has a common focus on evolutionary forces. But why do we have just one word? and is the underlying common feature of evolution a sufficient explanation for our monism? There is certainly good theoretical and empirical evidence that evolutionary groups can come into existence by a variety of ways and so too that they may persist in a variety of ways. In other fields, such a multiplicity of processes usually leads to a multiplicity of words. Researchers of diseases, or of storms, or personalities, to mention just a few examples, all find use for a single all-encompassing term (i.e., DISEASE, STORM, and PERSONALITY) and also make frequent use of subcategories. For the most part this is not so for SPECIES. Given how closely we study species, it is interesting that we do not tend to subdivide the species category into subkinds on the basis of different features of existence. Various words for sub-kinds of species have been designed and promoted, but most see little usage (chapter 1).

To see the absence of common terms, where one or more would be useful, consider our understanding that evolutionary groups in nature are often not distinct. Could we not make good use of a term for referring to species that are partly distinct, but also partly indistinct by virtue of exchanging genes with other species? We do get by, mostly by relying on modifiers of SPECIES. For example, we see increasing use of the adjectives MOSAIC and RETICULATE to describe species that are engaged in gene flow with other species (Arnold 1997; Harrison and Rand 1989).

There have been a number of proposals that multiple words for different kinds of species be used, but almost without exception these proposals are intended, not to distinguish different modes of species existence, but rather to facilitate the different ways that species are perceived. Thus, Ravin (1963), Gilmour (the DEME- terminology) (1939, 1954) and de Quieroz and Donoghue (1988) provide suggestions for a plurality of ways to refer to species, but in each of these cases the different criteria are based on ways that the investigator identifies or envisions species. They do not distinguish species on the basis of the processes that cause or maintain them as entities in nature.

I think the reason for our monism is that the bulk of our usage of SPECIES is not motivated by an understanding of the processes that cause species. Rather, it is caused by the way we construct categories of organisms, and thus it is ultimately motivated by recurrence among organism. We devise categories of organisms on the basis of similarity among them, and we use the symbol SPECIES to refer to those categories. Our intellectual understanding of real species, of entities in the world, does not change the fact that our perceptions of similarity have lead us to recognize real species, and that we must make use of that sense of similarity to identify real species. The one thing that all real species share, along with all of our categories that we equate with real species, is similarity among their member organisms. In this way, all species appear to us in the same manner, and it is this commonality that drives our monism. It is yet another way that our categorical imperative distracts us from understanding species, for it would be nice to develop a richer lexicon about the real entities that we study.

What Are Taxa?

Let us now turn away from species and take up questions regarding those named groups of organisms (taxa) that include multiple species, the so-called *higher* taxa. In particular, let us consider taxa that correspond closely to clades on well studied evolutionary trees. Such taxa fit the definition of MONOPHYLETIC GROUP, which is a group that includes not more than, and not less than, all of the organisms that descended from an ancestral species (chapter 10). Monophyletic taxa have a simple evolutionary justification, as both they and their defining characteristics can be easily mapped onto a phylogenetic tree diagram. This is not to say that tree diagrams are necessarily an accurate way to represent evolutionary history (often they are not—chapter 10), nor does it deny that there are other ways to devise taxa. But in what follows I inquire about how higher taxa might exist, and this question is difficult for taxa generally. It turns out to be a little less difficult for monophyletic taxa.

Recall the realist/nominalist debate over whether natural kinds do or do not have some existence out in the world (chapter 4). My pragmatic stance falls on the nominalist side of this debate, for I maintained that it is best to treat as real just those things that have some tangibility, those things that have some localization in space and time and have potential for being acted on. This stance has been a useful guide for considering the ways that evolutionary groups may exist as entities in the world. However that stance is not going to help very much in considering the existence of higher taxa. Unlike references to species, references to more inclusive taxa generally carry little or no connotation of individuality. Just as with other natural kinds, the members of a higher taxon are recurrent of one another, and we define them on the basis of shared properties. And like all natural kinds, higher taxa exist in our minds and in our communications. Thus, for example, *Class Mammalia* is one of our categories, but it is not an individual, at least in the sense that individuality has been described for evolutionary groups. All mammals do not form an existing tangible entity of interacting, or evolving, organisms.

But surely our mammal category, that includes all organisms with hair and mammary glands, pertains to *something* that is real out in nature. Saying otherwise would be strongly nominalist, and I guess that few biologists are interested in it. Biologists recognize an evolutionary explanation for the shared traits of mammals, and they recognize a direct historical connection among all mammals. In fact, all people are generally and deeply committed to the ontology of taxa, and this book has no purpose, nor means, for doubting that. This commitment is seen most simply in the words of simple sentences, such as "Today we will learn about mammals." One just does not say such things without implying that the mammal category exists in the world somehow (Quine 1961).

In reprising the realism/nominalism debate in chapter 4, I reiterated a nominalist argument for the individuality (and existence) of evolutionary groups. In this light, higher taxa, not being individuals, do not exist as such. But higher taxa do seem to be special, and biologists and laypersons treat them as real. If we are not to ignore this and are not to treat peoples' behavior regarding taxa as delusions, then we must admit some sort of realist stance about higher taxa. In short, we find an ontological puzzle, and we find ourselves getting enmeshed in questions of realism—a complicated, nay, messy area of philosophy (Hacking 1983). Dare we tackle it? It would be brave to do so, but pragmatism, like discretion, can be the better part of valor. I think we can make some good progress, in limited space, on how to think about taxa, without engaging too much the intractable, purely ontological question, "Do higher taxa exist?" As others have, philosophers and systematists alike, we shall suppose that they do, and instead focus mostly on the more practical and accessible question: "How do taxa exist?"

We can begin by considering a taxon's most accessible mode of existence, that has been emphasized repeatedly in this book, which is as a category in our minds and communications. Taxa, like other natural kinds, are devised because we find recurrence among our observations, and so in asking "How do taxa exist?" let us focus on what is recurrent and on what causes that recurrence. For monophyletic taxa, we can say that the recurrence has been caused by a shared evolutionary history, and that those things that are recurrent are the particular suites of unique traits (in the genotypes or the phenotypes) that appear again and again in the organisms that we study. We find, for example, that some organisms possess hair, and we also find that these same organisms possess the singular adaptation of mammary glands. These traits of hair and mammary glands always seem to co-occur, and this particular pattern of recurrence—of co-occurring, seemingly uniquely derived traits—has lead us to devise the mammal taxon.

So, apart from being a mental category, how does a taxon exist? The simplest answer, that follows from the way that we devise taxa, is that it exists as a recurrent pattern among the features of organisms that have been studied. Again and again we find a recurrent pattern in which many organisms are found to share particular unique features; and in a great many of these cases we can offer a straightforward evolutionary explanation (i.e., the pattern arose first in a common ancestral species). Of course, this does not always work. Organisms that share several derived traits may have had a far more complicated history than allowed for in a bifurcating tree model (chapter 10), and even if a branching model does fit, actually coming up with the best tree model can be difficult. But these complications do not invalidate the use of patterns that seem to consist of shared, derived traits, for the purpose of devising taxa. The patterns have been caused by evolution, and it is fair to treat them as real. The suggestion then is that the taxa that we devise and that exist in our minds are caused by patterns that exist in the world, and that it is fair use to identify those categories with those patterns.

In saying that taxa exist as patterns in the world, we need admit no confusion with the existence of evolutionary groups. An evolutionary group is an entity of interacting organisms whereas a taxon is a pattern found among the characteristics of organisms. An evolutionary group includes organisms and may include smaller evolutionary groups including entities that we might otherwise identify as families, or colonies or populations. Taxa are patterns, and as such do not include populations or any other kind of evolutionary group, and they include organisms only in the sense that such organisms manifest the pattern.

Quine, among others, has argued for a particular ontological status of natural kinds, while maintaining that such reality is different from the existence of entities (see, e.g., Gibson 1997). Dennett addressed the reality of patterns, in particular, and advocated a "mild realism" based on utility and mathematical compression. If a pattern is useful to one who recognizes it, say for the purposes of prediction, then it is real. So too if a pattern can be effectively summed up, or recreated, using a terse statement or algorithm, then it is real (Dennett 1991). Haugeland found Dennett's ontological position to be muddled (1993). In return, Dennett largely conceded this failing, and then simply refused to be drawn farther in to the ontological questions raised by his stance: "But must I attempt to put my ontological house in order before proceeding further?" (Dennett 1993, 213). For Dennett, the obvious answer to the rhetorical question is "no"—he will not do ontology, no matter how much he may seem to be doing ontology. This is awkward, but it is pretty much the stance that I am taking here. Higher taxa are eminently useful, and so it is a good idea to treat them as real, and we'll not say much more about the ontology than that. If a category or an idea in the mind turns out to be a good one for making sense of the world, then this can be taken as evidence that the category actually matches up with something in the world. Like Dennett, I find it easier to write about evidence of existence, than to justify a firm stance on whether something definitely does, or does not, exist.

We must note, however, that there is one thing that sets higher taxa apart from other natural kinds. For higher taxa like mammals and birds, but not for most other natural kinds like comets or ocean waves or toothaches, recurrence has been inherited. In the case of a monophyletic taxon, a distinct, finite series of events during the time of common ancestry has been passed on, in the genes, to all members. The members of a monophyletic taxon share features that are not simply similar, nor analogous, nor convergent, as might be said of the shared features of the members of some other natural kind. Among the members of a higher taxon, the key shared features are literally homologous by virtue of DNA replication. Do we envision any such direct historical continuity among the members of other natural kinds (e.g., among all comets, or all ocean waves, or all toothaches)? Even if we do envision some common historical threads among disparate members of other natural kinds, it is still the case that monophyletic taxa are historical standouts. The recurrence among the members of a higher taxon is caused by their all sharing information (literally, in their DNA sequences) that they inherited from common ancestral DNA sequences.

So monophyletic taxa are special natural kinds. Does their historical connectedness entail some existence, out in the world beyond our minds? It does indirectly, for it is because of shared history that monophyletic taxa are useful. However beyond the utilitarian argument, the fact of shared history does not help us better address the ontology; for if we grant that existence is a contemporaneous property, then we cannot recognize it for historical reasons alone. As stressed above, utility, and not ontological standing, is the key to our appreciation of higher taxa. It is what helps us choose among patterns and among kinds of patterns, and it is the criterion that motivates an emphasis on monophyletic taxa.

As a practical matter, monophyletic taxa do not have some special status other than utility. If monophyletic taxa are real, then so too are there are an effectively infinite number of other real patterns out there, caused by evolution and ancestry, that are *not* monophyletic, and that could still beget categories. This has, of course, occurred countless times, for many of our taxa are useful and are not monophyletic. Consider the oft-described example of the *Class Reptilia*, which, like most taxa, can be identified as a pattern of traits that are shared among some organisms due to evolution and common ancestry. Reptiles (including snakes, lizards, turtles crocodiles, and tuataras) are air-breathing, cold-blooded vertebrates, with internal fertilization, and scaly skin. However, these features are a mixture of old traits, that are not unique to reptiles (amphibians are also cold-blooded, air-breathing vertebrates), and new traits, that have not all been passed on to other organisms that also descended from the common ancestor of reptiles (birds are also descendants of the same common ancestor from which sprang reptiles, and they do not have scales and they are not cold-blooded). *Class Reptilia* is widely used (and is thus useful) but it is not monophyletic. I would agree with those who argue for replacing our primarily Linnaean taxonomic system, with one that placed a primacy on monophyletic taxa. The current system is full of useful taxa, but in the long run, a revised system that focused on monophyly would be more useful.

Species as categories (i.e., not the large evolutionary groups) are taxa and they certainly have their uses. In the next chapter, I address their utility and how best we might refer to them. But here, in an explicit discussion on the reality and utility of taxa, I have ignored species taxa. This discussion of taxa has focused strongly on monophyletic groups, and as discussed at length in chapter 10, species taxa are frequently not monophyletic. In this regard, it is telling that de Queiroz's and Gauthier's proposal for revising current taxonomies, with a system based on monophyletic taxa, is deliberately directed at higher taxa only, and avoids any consideration of how to identify species taxa (de Queiroz 1996; de Queiroz and Gauthier 1994).

Nominalism of Taxonomic Levels

Today there is no truly nominalist voice in the systematics debates. So far as I know, no one is arguing for the nonexistence of all taxa. But there does persist a somewhat related idea that denies that the species category (the category of all species taxa) is a unique and special one among all categories of taxa. When Darwin wrote "how entirely vague and arbitrary is the distinction between species and varieties" (Darwin [1859] 1964, 48), he was clearly saying that species are indistinct in a particular way, that is, that the species category is not distinct from the variety category. Darwin repeatedly made reference to species with explicit or implicit connotations of reality. Our main clue to this is that he wrote of species as if they could do things, or have things done to them. Darwin wrote of species as if they have beginnings and endings and as if they evolve. But he did make a good case for a kind of nominalism, something we might call LEVEL NOMINALISM, though it is quite different from a strong philosophical nominalism. Level nominalism is the idea that we cannot tell where to draw the line over the degree of taxonomic distinction and thus do not know what rank or level (e.g., Genus, Family, Species, etc.) to apply. Thus, for example, when arguing as a level nominalist, Nelson said that SPECIES is the name of a taxonomic category, and that instances of that category (i.e., particular species) are taxa, but insisted that these distinctions are not different from, and not less arbitrary than, decisions over more inclusive taxonomic categories. For Nelson, the distinctions made when assigning taxonomic rank are entirely arbitrary, but nevertheless, "All taxa are real" (Nelson 1989). It is like saying that the distinctions between rivers, streams, creeks and rivulets are arbitrary, but that all channels of flowing water are real. Note that if one embraces this idea, then species are also not terminal, or basal, taxa, that special category of organismal kinds whose members reside at the tips of evolutionary trees. Level nominalism requires embracing the fractal view of evolutionary hierarchies. And just as one could not count all channels of flowing water, a level nominalist could not count real taxa. The reason is that one never knows how small an apparent entity must be in order to count it (chapter 6).

It should be stressed that biologists generally recognize that the level designations (Genus, Family, Class, etc., any taxonomic rank apart from the species rank) are inherently arbitrary. The analogy to channels of flowing water is quite close, as a great many taxa are commonly taken to be real in some sense, and created by evolution over history, but whether we identify such a taxon to be a Class or an Order is a practical matter of convenience and convention. The SPECIES category is another matter, and for many systematists, the search for the best species concept is very much a search for just what it is that makes the SPECIES category different from other taxonomic ranks. There are no debates of remotely similar scope for those other categories.

Some have noticed that Darwin's claim of indistinct species does not seem consistent with his many references to them that suggest they are real. Consequently, there has been some debate about what he really meant. Was he posturing—imperfectly, but as a kind of philosophical nominalist—for the sake of argument when he said that species of organisms are not real (Beatty 1985), or did he just simply believe that the species category was not real, as would a level nominalist, but that organisms still occur within real evolving groups? The latter is almost certainly closer to Darwin's meaning (Stamos 1996).

FOURTEEN

WHAT IS TO BE DONE?

One agreeable aspect of the explanation of the species problem is that it does not entail a radical or utopian corrective. Biologists have figured out some good ways to study species, notwithstanding a fair bit of confusion and miscommunication, and so it is not surprising that many of our traditions of studying species and referring to species remain fairly well justified even after we come to an understanding of the species problem. For example, it is clear that many definitions of SPECIES are roughly synonymous with the idea of large evolutionary groups, and so necessarily I do not think that we have been off base when we think of real species in these ways.

Another tradition that still stands in the light of a species problem explanation is the phylogenetic systematics of higher taxa. Taxa are certainly categories, but I think that some are more than that. Because of the strong historical connectedness among the organismal members of higher taxa, and especially because of the utility of higher taxa, it is appropriate to treat them as real patterns in nature. I concur with those phylogenetic systematists who would place a premium on trying to find patterns that suggest monophyly of the organisms that would be included in higher taxa. But keep in mind that even with some ontological acknowledgment, higher taxa have a mode of existence utterly different from entities such as evolutionary groups.

Still, the species problem hinders research in evolutionary biology and in systematics. With the understanding that our own conflicting motives have caused the problem, we can realize some ways to do better, and we can see how some species-related traditions should be changed or scrapped. One thing we would do well to abandon is much of our debate over definitions of SPECIES. Most philosophers and biologists who have participated in the lexical squabble have been confused about categories, and many have long missed the basic points that short definitions are poor containers of knowledge, particularly if they are meant to describe and arbitrate a fuzzy reality. Perhaps it was understandable and reasonable that we would wish or assume that our categories match up well with real evolutionary groups. But reasonableness aside, we now see that assumption as a burden that has driven the SPECIES-definition industry. I would encourage readers and understanders of the argument in this book to not approve of a continuing debate. For those who would persist in quarrelling over the best terse definition for describing real species, the argument of this book provides an explanation of their persistence, which is that they follow a categorical compulsion and that this distracts them from their goal of understanding.

One area where definitions can continue to play a constructive role in our discourse about species is the delineation of kinds of real species. As discussed in chapter 13, biologists have a strong tendency toward species monism that is probably a byproduct of our confusion over categories. If we can move beyond that confusion, then old and new definitions of kinds of real species will be more useful.

Taxonomic Species

It must be emphasized that it would be counterproductive and futile to try to stop referring to organisms in a categorical way, or to stop giving names to kinds of organisms. Even if some organisms constitute an evolutionary group they will first appear to us as recurrent aspects of nature. More than a tradition, our categorical references to kinds of organisms are a necessity. There is simply no other avenue open to us, for considering organisms as parts of entities, that does not first pass through having treated them as a category. This is because the starting point for thinking about species, and the most basic evidence for consideration that organisms are part of an evolutionary group, is some kind of similarity. It does not matter if that is some sort of overall similarity, or if it is some sharing of seemingly derived character states, as envisioned by cladists. Either way we are confronted with recurrence, and our language in regard to those organisms must pivot around that recurrence. We must devise a category and give it a name. We do this regularly, most often fairly informally, such as when campers complain about the "little black bugs." Biologists must also be more formal, and when they recognize recurrent organisms that do not fit in previously recognized categories, they must describe that recurrence and label the category. These formal designations also require decisions about where the newly recognized category is to be placed among existing taxa within the existing taxonomic system. The task then, for those who would wish to understand evolutionary groups and for those who would wish to refer to real evolutionary groups, is to not mistake the taxa for the entities. Perhaps biologists should have a litany to help keep things straight; maybe something like "Do not mistake categories for entities; do not mistake taxa for real species; taxa are mental categories that correspond to patterns in the world, and they are not evolutionary groups."

Cautions like these and having an understanding of the causes of the species problem may help biologists avoid perpetuating some aspects of the species problem. However they do not remove one major area of uncertainty, which is how to identify and use species taxa. Species taxa are those named categories to which we have given the taxonomic rank of SPECIES. In the last few centuries, biologists have created nearly 2 million such categories and for only a fraction of these do we have any insight on how, and to what degree, the living organisms that we might include in the category actually constitute evolutionary groups. Our species taxa have been devised on the basis of recurrent patterns (just like higher taxa), but they are widely taken to be patterns of recent evolutionary origin and also to be phylogenetically basal.

Species taxa are very problematic, for they are more difficult to devise than higher taxa, and their correspondence to real species is usually unknown and often nearly unknowable (chapter 12, and see below). Unlike higher taxa, species taxa cannot generally be devised on the basis of monophyly, so it is not clear what criteria should be selected as their defining characteristics. Add to this difficulty the heavy traditional pressure to use the species category only for basal taxa, and it is not hard to appreciate why systematists who would apply the species category are constantly wondering where to draw the line in such a way that someone else does not later say that the taxon they have devised is not basal. Nor will the existence of real species—real evolutionary groups —be much help. Systematists considering the creation of a species taxon will usually have little knowledge of the degree to which their supposed taxon corresponds to evolutionary groups. Nature is little help to categorizers and generators of species taxa, for they must build something fine and agreeable to others, and they must do so using indistinct raw materials.

In sum, the difficulty with species taxa is that we do not know by what rules to devise them—how much novelty and how much recurrence justifies a new species taxon?—and we don't know how they correspond to real evolutionary groups. These difficulties are likely to stay with us. Biologists, systematists, and evolutionary biologists alike have had the hope that species taxa would match real distinct evolutionary groups in nature (chapters 11 and 12). This is just not going to happen for very many species taxa, and of those times when it does, the evolutionary group will often be diffuse and indistinct in parts. Real species are difficult to identify because of the species-sampling problem (see below), and they are often indistinct and nested within one another. Even in those cases when we do, with much research, match a taxon with an entity, it will often be the case that the entity is far less distinct than the taxon is. For example, many species that seem distinct in some ways, also exchange genes with other species (chapter 7).

Matching Up Real Species and Species Taxa

These difficulties do lead to a suggestion, though in most ways it is an affirmation of some things that careful biologists already do. If one goal is to understand the correspondence between species taxa and real species, and to devise species taxa in such a way as to have some such correspondence, then a good starting point is to include as much geographic, ecological, and behavioral information as possible when describing new species taxa (Winston 1999). Simple phenotypic or genotypic similarity is limited evidence of a common evolutionary group, but the more evidence that can be brought to bear, and to the extent that evidence is consistent with a common evolutionary group, then species taxa are well motivated. Such taxa reflect a pattern, but we may *hypothesize* that all such living organisms that fit that pattern actually constitute an evolutionary group.

Evolutionary biologists and systematists make hypotheses about the existence of real species all the time. A ubiquitous biological protocol begins as follows: Organisms are collected or observed in the field; among those organisms, some are found that resemble one another but are novel with respect to other described forms; their disparity from known forms, and their recurrence of each other are taken as evidence of a heretofore undescribed species; and a species taxon is devised, named, and described. So far so good. At the end of this sequence we have a species taxon, a category based on a pattern found in nature.

Now could it also be the case that organisms in this category also constitute an evolutionary group? It could, but answering this question requires investigation of the forces that occur in evolutionary groups, and
that can be very difficult. Suppose, however, that we undertake this investigation and that we find that all the organisms that we sample and
that fit the taxon also show evidence of composing an evolutionary
group. Such evidence could come from direct observation, or from genetic methods, or from any method that is sensitive to the presence of
shared evolutionary forces. If the evidence is consistent and ample, then
again, so far so good; but now here is the tricky part. Suppose that at a
later time, another sample is taken, and the same pattern is observed as
lead to the original species taxon. Could these organisms also be taken
as components of that same evolutionary group as was discovered for
the original species taxon? They cannot, at least not without risking
error. Any and all conclusions that these newly considered organisms
are members of the same evolutionary group, as were found in the previous sample, are inductions; and thus, they are tantamount to equating
the category with the entity. Consistency can however be maintained
by proposing the hypothesis that the newly considered organisms are
part of an evolutionary group that has descended from the one that was
previously revealed.

The idea that a named category can correspond to a real species is
equivalent to the hypothesis that the recurrence that inspired the category, as well as *all additional recurrence* that might ever be considered to
fall within that same category, was the result of all those recurrent organisms collectively comprising an evolutionary group. Such an extension of a category, to a possibly large number of things not yet seen,
must be a hefty induction. It may seem reasonable because it is combined with a mechanistic justification, which is just that evolutionary
groups are known to be the major cause of recurrence among organisms. One major catch with such inductions is that recurrence can long
outlive the circumstances that caused it. However, the real problem
with allowing a category to serve as a model of an evolutionary group
is that the induction—the hypothesis that all things in the category will
also be in a single species—is not easily tested. This is the species-
sampling problem (see chapter 12 and below). Many authors have written of how species taxa can serve as hypotheses of real species, and they
are exactly correct (see e.g., Baum and Donoghue 1995; Mallet 1995;
Mayr et al. 1953; Sites and Crandall 1997; Templeton 1994). But while
many biologists are engaged in the detailed study of one or a few real

species, in practice very few have the time or energy to research and examine all of the categorical species they use.

The suggestion then is that we use SPECIES TAXON or TAXONOMIC SPECIES to refer to a named kind of organism that is based upon a unique pattern of biological diversity that we think has been caused by evolution; and that we use SPECIES, or REAL SPECIES, to refer to a large evolutionary group. Furthermore, we must recognize the difficulty of identifying evolutionary groups, and so an instance of SPECIES should be motivated by an hypothesis of an evolutionary group, rather than the assumption of one. Many biologists already do these things to some degree. But even for those who are in the habit, the species-sampling problem and our tendency to make large inductions regarding species mean that these recommendations will be difficult to follow.

It is best to be frank about the difficulties of examining the correspondence between species taxa and real evolutionary groups. Since we *must* generalize from what we find in our samples, evolutionary biologists are constantly being thrown into the category/entity muddle. Furthermore, it is not just a philosophical or semantic muddle. Our inductions that are based on recurrence are often wrong for the simple reason that recurrence is often a poor guide to real species. This point bears reiteration to head off what is likely to be a pressing question, which is "How many of our taxonomic species actually do correspond to distinct evolutionary groups?" Note that any well-read naturalist could quickly list many cases where a species taxon is associated with some evidence that representatives constitute an evolutionary group. The problem, however, in most cases that indicate some correspondence, is an absence of evidence on the distinction of an evolutionary group or on how many other such groups it may contain. Species taxa are categories based on seemingly distinct patterns, and evolutionary groups are fuzzy and nested. Given the difficulty of studying real evolutionary groups, and given the growing evidence that such groups are not distinct and often do not match up well with taxonomic species, it is likely that most present-day attempts to answer the question would make the same mistake that Mayr made when he attempted to assess the correspondence between plant species taxa and biological species under the BSC (Mayr 1992); see chapter 12. The processes that make real species are hard to study, and it is all too easy to fall back on a simpler categorical view of species.

Counting Species

Many people, biologists especially, enjoy counting species; and all manner of ecological and evolutionary research (both empirical and theoretical) makes use of species counts. With very few exceptions, claims on the magnitude of biological diversity or of a comparison of such magnitudes are cast in terms of species numbers. For real, well-studied species that are large evolutionary groups, it may sometimes be reasonable to employ integers for quantification. Sometimes these entities are going to turn out to be distinct enough that counting is a good way to quantify, much as we sometimes find clouds to be countable. But even distinct large evolutionary groups will sometimes contain smaller evolutionary groups that seem fairly distinct, and when this occurs counting becomes problematic. Nobody could ever count all the evolutionary groups, and this is for a reason that has nothing to do with large numbers or with the difficulty of perceiving evolutionary groups. More simply and fundamentally, it is because there do not exist a certain number of them—in the same way as for clouds and populations. Also, because of the species-sampling problem, real species are very difficult to study with sufficient thoroughness to know how distinct they are and to what degree they contain smaller evolutionary groups. If and when real species have been studied so well that we understand their structure and distinction, it will probably also be the case that mere counts will seem like the least interesting way of quantifying them.

Now consider taxonomic species, and let us ask how these might be counted. Keeping in mind that taxa exist as both human categories *and* real world patterns, what might be the major determinants of the number of taxa? Quite generally, there are two major factors: the human component, which includes all of our capacities for sensory perception, cognition, tool use, sample collecting, and communication; and the external component, which includes all the real recurrent patterns that inspire our human categories. For brevity in the remainder of this discussion on counting taxa, we can call the human part RECOGNITION and the external part PATTERN. The distinction allows us to think about when and how the counting of taxa might be a repeatable process and thus how counting might be useful.

But wait. If a large part of a species taxon is the human recognition apparatus, then aren't taxonomic species counts entirely meaningless and useless? The answer might seem to be "yes," particularly if one understands how complex and variable the human recognition process might be. As I will explain, I do think it is folly to suppose we might have consistent taxon counts on an absolute scale. But there are contexts where we can have consistency and where counts of taxa, including species taxa, are useful. The reason for this is that humans do not vary willy-nilly in the ways that they interpret nature. We do share common abilities and levels of skill in all of the components of recognition, and we communicate about the things we recognize. If people worked together to use common methods and common levels of effort, taxon counts could become stable among investigators. No doubt it would be a very large common effort to agree to common methods, and such methods would necessarily be largely arbitrary, at least with respect to real species. But these difficulties are not new ones, and systematists have been working to overcome them for generations.

The genuine problem with taxon counts, even repeatable ones that are arrived at with a consensus on methods, is that we don't know just what they are counts of. It does not help very much to think that taxa are also real patterns. As true as that may be, there will be an effectively infinite number of patterns that we might recognize and that might seem useful. As the history of species taxonomy well shows, a change in the effort or mode of taxon recognition leads to a change in the count. There is also no escaping the fact that each and every taxon, and thus each and every count, is the result of an interaction between real patterns and the human mode of recognition. Knowing this, we must appreciate that any particular count of species taxa, considered out of context and without information on criteria and effort, is meaningless. What systematists can do is have repeatable species counts, and they can do that by agreeing to, and holding to, guidelines for the creation of taxonomic species and by holding to precedence for those species taxa already described.

Of what use is a count that is consistent among biologists, but that actually enumerates some idiosyncratic interaction between patterns of biological diversity and processes of human recognition? It certainly cannot be taken as an absolute measure of patterns in nature. However, it can be used to assess changes in those patterns. If we have consensus on the methods of identifying taxa, and we can agree on taxon counts, then we will also be able to recognize when the count is changing because of changes in patterns in the world. This goes on all the time in particular geographic contexts or for particular higher taxa, as biologists regularly report changes in the numbers of taxa. Typically such counts are decreasing because organisms are dying and patterns among living organisms are disappearing (extinction). Thus, even though the absolute counts in these contexts may not have the meaning people often assume of them, the changes in such counts actually do reflect changes in the amount of biological diversity.

An appreciation that taxa have two components does entail a shift in our thinking about the way that we think taxa are representative of nature. In chapter 11, I referred to the common aversion for the notion that "a species is just what a systematist says it is." But because of the large human component of our taxa, there is a related idea that I think we must prepare ourselves to accept, and that is that a species taxon is what systematists agree it is (Heywood 1998). This is particularly true for those many times when such designations will seem palpably arbitrary. It will generally be the case that there is little people can do to dispel our own role in taxon recognition and thus the arbitrariness of our taxa. Whatever consensus we generate will come via common recognition of the subjectivity of our species taxa and of our need for species taxa. The exceptions will be those few cases when the correspondence between taxa and evolutionary groups have been well researched. In such cases, species taxa may bear some additional affirmation above and beyond that conferred by systematists' consensus. Most species taxa today have not been studied much beyond their existence as a recurrent pattern, and so they simply are little more than named recurrent patterns. Importantly, such little studied species taxa can still be very useful. We should not abandon them or stop generating them, but we should recognize them for what they are.

Another shift entailed by an appreciation of the human component of taxa is a giving up of the idea that we shall have absolute counts of all real patterns in nature, particularly all real taxonomic species. It is common for biologists to report absolute global counts. At one extreme are literal counts of all named species taxa. For example, the editors and authors of the Global Biodiversity Assessment of 1995 brought together diverse experts who carefully tabulated the number of distinct, described species (Heywood and Watson 1995). They came up with a total of 1.75 million reported species worldwide (Hawksworth et al. 1995). More common than this sort of literal count, are hypothetical counts. It is common today to see estimates of the number of undescribed species that are really out there; and many wide ranging, very large numbers are now in the public domain. Consider a quote from a Congressional Research Service brief to the United States Congress: "Current estimates of total species range from 5 million to 100 million, with 10–30 million being commonly accepted numbers" (Corn 2000).

The broad human interest in the quantity—we can call it S—of the number of real species, out in nature, is quite a phenomenon. The S notion, with or without actual numbers plugged in, is an integral part of a great deal of discourse on biological diversity, among professional biologists and laypersons alike. It is a mountain of a tradition and will probably not be shifted any time soon, even though S is a chimera.

Biologists are right to try for consensus on methods for recognizing taxa, and for counting the taxa they recognize, for such numbers can be used to track changes and to make comparisons among different geographical regions. But the sooner we recognize that taxa lie partly within us, and the sooner we stop dwelling on S and other hypothetical absolute counts of species taxa, the better scientists we will be. Real evolutionary processes have given rise to unknown, albeit vast, numbers of DNAs, cells, and organisms, and our growing familiarity with their diversity could lead us to recognize an unending, effectively infinite, tally of patterns among them. It is silly to ponder just how many species taxa we might generate if we could see all of biodiversity.

Preserving Biological Diversity

The human species is changing our planet at an ever increasing rate and in the process other evolutionary groups are rapidly dying off and many patterns of biological diversity that we might recognize, and use to devise taxa, are disappearing. Unfortunately, our species problem partly gets in the way of our attempts to quantity and stem this loss of diversity. The problem is that the standard metric for lost biological diversity is a change in S. The effects of altered environments, particularly the loss of habitat, are regularly cast in terms of the loss of unknown real species—our unknown S is getting smaller. Government agency reports and newspaper articles, written either by professional biologists or with their input, regularly remark on the loss of biological diversity in terms of the decreasing number of real species. Of course, these reports necessarily entail a full dose of the species problem. From all that has been said, we must recognize that virtually all statements like "X species have been lost" or "species are being lost at a rate of X per day" are ambiguous in ways that their authors did not intend. First, such statements are almost always some kind of extrapolation to absolute counts of as yet unidentified species, and are thus part of the chimeric S. Second, the numbers are reported as if they are counts of real species and they are reported without any insight or acknowledgment of the confusion between species taxa and real species. Third, even when such reports pertain only to already recognized taxonomic species, the absence of consensus among systematists means that measurements and reports done by others will return different numbers. We can also trust that the biologists who report such extinction counts know something of their inherent nonsense, for systematists and evolutionary biologists are experienced in species count inconsistencies (chapter 2). Is there some reason why the practice of reporting counts of lost real, unknown species should not be seen as a foisting of the worst of biologists' inconsistent and unscientific species numerology? In their concern for what is being lost, biologists overlook their uncertainty about SPECIES (Rojas 1992). Surely we can do better.

One thing that would help in these matters is if lost diversity was reported without the taxon/entity ambiguity and if, instead, counts were described clearly in terms of taxonomic species, rather than real species. The reason is simply that taxonomic species are vastly more accessible and enumerable than real species. Still, there are two problems with this approach. First, it would work much better if there were consensus on the methods of identifying species taxa and on the criteria of recognized species taxa, and such broad consensus does not yet exist among systematists. Again, such consensus must necessarily continue to be based on criteria that have much to do with methods of pattern recognition and with human discourse, and that are partly arbitrary with respect to nature. Second, given the funding rate of systematics research, the pace of systematics research and of pattern discovery is slow. As things stand, only 10,000–20,000 new species taxa are recognized each year, and it is essentially impossible for most biologists to know of and make use of the recognition criteria that have been devised for most species taxa. Furthermore, it will be some time, perhaps a long time, before there exists a broad consensus on the methods and criteria for devising species taxa. In short, systematics is not going to "catch up" and somehow devise a thorough species taxa catalog of biological diversity in time so that it may be used to help stem the loss of biological diversity.

There are other ways to assess and to preserve biological diversity that do not presume knowledge of real species or of unambiguous species taxa. One way is to rely on the hierarchical taxonomic system and to work to preserve higher taxa that are more inclusive than species taxa. The case can even be made that, by taking care of more inclusive taxa, many of the species taxa that would be categorized within those taxa will also be taken care of (Williams and Gaston 1994; Williams et al. 1997). Such a method would necessarily suffer the uncertainties of hierarchical taxonomies as models of phylogenetic history (chapter 10) as well as the inconsistencies in methods of application of taxon rank among systematists, but to the extent that such a model is approximately correct then it may well serve (Williams et al. 1997). The shortcoming of this approach is the risk that we will lose organisms representing as yet undescribed patterns that have not yet been placed in higher taxa or have not yet been placed in the appropriate higher taxa, given their evolutionary history. The ongoing discovery of novel organisms and continuing systematics research are the only ways to overcome this problem.

A variant of the method of preserving all taxa, of some rank, is to preserve just those that are the most desirable, as assessed by some criteria of desirability. The value of a taxon may depend on its utility to people, or its ecological distinctiveness, or on how evolutionary unique it appears to be (Humphries et al. 1995; Vane-Wright 1996; Vane-Wright et al. 1991; Warwick and Clarke 1998). None of these qualities are easy to assess, and the general approach does not make any explicit acknowledgment or correction for the species problem. But again, the methods can be applied to taxa that are more inclusive than species and by so doing they can avoid species uncertainty.

An entirely different approach, for assessing biological diversity while avoiding the species problem, is to ignore taxa, and to numerically calculate diversity by some measure other than counts of taxa. For example, one could measure the amount of variation found in the DNA genomes of a sample of organisms that had been sampled randomly from nature, and do so without regard to their species or other taxonomic designations. Such a sampling scheme might seem strange, and it would probably be unpopular as a general method, but it is exactly what is done when small organisms are sampled from extreme environments that people rarely observe. For many of the organisms that live in sediments or soils, or in extreme places like underground oil fields, biologists have few preconceptions about species, and they find that the most direct way to assess biological diversity is to isolate DNA directly from samples containing many microorganisms (Embley and Stackebrandt 1997; L'Haridon et al. 1995; Munson et al. 1997; Vetriani et al. 1998). Then by measuring the amount of DNA sequence variation in their samples, they can quantify the diversity. Such measures are uncontaminated by the species problem, though they do raise other difficult questions regarding sampling protocols.

Taking Care

We are as one with our need for categories, and we are not omnipotent observers, and so we must endure the species-sampling problem. Our best recourse is to keep in mind the ways that our conclusions regarding real species, which we obtain from our samples, may lead us to wrongly generalize about the things not in our samples. Of course, this is also consistent with maintaining the mindset that taxonomic species can serve as hypotheses of real species.

The primary recommendation of this book is that biologists pay closer attention to their organismal kinds and their usage of SPECIES. Categories, especially kinds of organisms, come so easily to us that we have grown up using them in ways that hinder our understanding of biological diversity. The need to overcome this tendency does not demand a revolution in the way we study or refer to species, but it does demand that we pay closer attention to our studies and our references. I would urge that biologists explain and justify those times they use SPECIES. They should make lucid statements that leave no doubt whether a species reference is intended to be a hypothesis of a real entity or whether it is to a taxonomic species. Finally, listeners, readers, and editors should not let them do otherwise. Knowledge does not come without responsibility. That is not intended so much as an ethical creed (though it is that) but rather as a practical one. For a scientist whose goal is understanding, pertinent knowledge cannot be steadily ignored without corrupting one's central pursuits. Our study of species and our discourse over species is partly dysfunctional, but by appreciating the causes of the species problem, we can do better.

Though the causes of the species problem are not to be simply undone, they can be overcome. The major clue that feeds this optimism is the way that the species problem has already been partially overcome. So much of our current understanding has come despite severe impediments. The rejection of typological, essentialist views of species is a fine example. Though we may still be confused by our categorical predilections, the widely accepted knowledge that these cannot be part of our best theories is no mean feat. So too are the modern discoveries of the complicated relationships that may occur between evolutionary trees of DNAs and the evolutionary histories of the organisms that contain those DNAs. In recent years, there has been a strong convergence between evolutionary biologists and systematists, in that both groups increasingly have research programs focused on the complex relationships between the histories of genes and the histories of groups of related organisms that contain those genes.

In the long run, I think the best that we could hope for regarding the species problem is that it someday be seen in the same way we see the water-breathing problem: Humans cannot breath water, and this absence of skill is a steady inconvenience and sometimes our undoing. But generally we have no confusion over water or our skills regarding it. We know that we cannot breath it, and so we have developed other ways to deal with water when we wish, or when we must. Such fixes are crude compared with those evolved by fishes, but we manage. Similarly, we cannot stop using categories as proxies for real species and as a step on the way to understanding real species, but if we are aware of the confusion that arises as a result, then we can compensate.

REFERENCES

Anderson, E. 1949. *Introgressive hybridization*. New York: Wiley.

Arnold, M. L. 1994. Natural hybridization and Louisiana irises. *Bioscience* 44:141–147.

Arnold, M. L. 1997. *Natural hybridization and evolution*. New York: Oxford University Press.

Atran, S. 1990. *Cognitive foundations of natural history*. Cambridge: Cambridge University Press.

Avise, J. C., and R. M. J. Ball. 1990. Gene genealogies and the coalescent process. Pp. 45–67 in *Oxford surveys in evolutionary biology*. Edited by P. H. Harvey and L. Partridge. New York: Oxford University Press.

Avise, J. C., and K. Wollenberg. 1997. Phylogenetics and the origin of species. *Proc. Natl. Acad. Sci. USA* 94:7748–7755.

Bachmann, K. 1998. Species as units of diversity: an outdated concept. *Theory in Biosciences* 117:213–230.

Barton, N. H. 1995. A general model for the evolution of recombination. *Genet. Res. Camb.* 65:123–144.

Barton, N. H., and B. Charlesworth. 1998. Why sex and recombination? *Science* 281:1986–1990.

Baum, D. A., and M. J. Donoghue. 1995. Choosing among alternative "phylogenetic" species concepts. *Syst. Bot.* 20:560–573.

Baum, D. A., and K. L. Shaw. 1995. Genealogical perspectives on the species problem. Pp. 289–303 in *Experimental and molecular approaches to plant biosystematics*. Edited by P. C. Hock and A. G. Stevenson. St. Louis: Missouri Botanical Garden.

Beatty, J. 1982. Classes and cladists. *Syst. Zool.* 31:25–34.

Beatty, J. 1985. Speaking of species: Darwin's strategy. Pp. 265–281 in *The Darwinian heritage*. Edited by D. Kohn. Princeton, NJ: Princeton University Press.

Berlin, B. 1992. *Ethnobiological classification*. Princeton, NJ: Princeton University Press.

Berlin, B. 1999. How a folkbotanical system can be both natural and comprehensive: One Maya Indian's view of the plant world. Pp. 71–89 in *Folkbiology*. Edited by D. L. Medin and S. Atran. Cambridge: MIT Press.

Berlin, B., D. E. Breedlove, and P. H. Raven. 1966. Folk taxonomies and biological classification. *Science* 154:273–275.

Briggs, D., and M. Block. 1981. An investigation into the use of the '-deme' terminology. *New Phytol.* 89:729–735.

Briggs, D., and S. M. Walters. 1969. *Plant variation and evolution,* 1st ed. London: Weidenfeld and Nicolson.

Briggs, D., and S. M. Walters. 1984. *Plant variation and evolution.* 2d ed. Cambridge: Cambridge University Press.

Briggs, D., and S. M. Walters. 1997. *Plant variation and evolution.* 3d ed. Cambridge: Cambridge University Press.

Brown, D. E. 1991. *Human universals.* New York: McGraw-Hill.

Brunner, P. C., M. R. Douglas, and L. Bernatchez. 1998. Microsatellite and mitochondrial DNA assessment of population structure and stocking effects in Arctic charr *Salvelinus alpinus* (Teleostei: Salmonidae) from central Alpine lakes. *Mol. Ecol.* 7:209–223.

Burma, B. H. 1954. Reality, existence, and classification: A discussion of the species problem. Pp. 193–209 in *Concepts of species*. Edited by C. N. Slobodchikoff. Stroudsburg, PA: Dowden, Hutchinson & Ross.

Butlin, R. 1989. Reinforcement of premating isolation. Pp. 158–179 in *Speciation and its consequences*. Edited by D. Otte and J. A. Endler. Sunderland, MA: Sinauer Associates.

Cain, A. J. 1954. *Animal species and their evolution.* London: Hutchinson.

Cantino, P. D., R. G. Olmstead, and S. J. Wagstaff. 1997. A comparison of phylogenetic nomenclature with the current system: A botanical case study. *Syst. Biol.* 46:313–331.

Caplan, A. L., H. T. Engelhardt, Jr., and J. J. McCartney (Eds.) 1981. *Concepts of health and disease.* Reading, MA: Addison-Wesley.

Chandler, C. R., and M. H. Gromko. 1989. On the relationship between species concepts and speciation processes. *Syst. Zool.* 38:116–125.

Claridge, M. F., H. A. Dawah, and M. R. Wilson. 1997a. Practical approaches to species concepts for living organisms. Pp. 1–15 in *Species: The units of biodiversity*. Edited by M. F. Claridge, H. A. Dawah, and M. R. Wilson. London: Chapman & Hall.

Claridge, M. F., H. A. Dawah, and M. R. Wilson. 1997b. *Species: The units of biodiversity.* London: Chapman & Hall.

Clark, J. B., and M. G. Kidwell. 1997. A phylogenetic perspective on P transposable element evolution in Drosophila. *Proc. Natl. Acad. Sci. USA* 94:11428–11433.

Clarke, B. C., M. S. Johnson, and J. Murray. 1996. Clines in the genetic distance between two species of island land snails: How "molecular leakage" can mislead us about speciation. *Phil. Trans. Roy. Soc. London Series B* 351:773–784.

Cohan, F. 1994. Genetic exchange and evolutionary divergence in prokaryotes. *Trans. Ecol. Evol.* 9:175–180.

Corn, M. L. 2000. Endangered species: continuing controversy, Online: http://www.cnie.org/nle/biodv-1.html. The National Council for Science and the Environment, Washington, DC. Accessed: March 29, 2001.

Coyne, J. A., and H. A. Orr. 1989. Patterns of speciation in Drosophila. *Evolution* 43:362–381.

Cracraft, J. 1989. Speciation and its ontology: The empirical consequences of alternative species concepts for understanding patterns and processes of differentiation. Pp. 28–59 in *Speciation and its consequences*. Edited by D. Otte and J. A. Endler. Sunderland, MA: Sinauer Associates.

Cracraft, J. 1997. Species concepts in systematics and conservation biology—an ornithological viewpoint. Pp. 325–339 in *Species: The units of biodiversity*. Edited by M. F. Claridge, H. A. Dawah, and M. R. Wilson. London: Chapman & Hall.

Crick, F. 1994. *The astonishing hypothesis: The scientific search for the soul*. New York: Scribner.

Crisp, M. D., and G. T. Chandler. 1996. Paraphyletic species. *Telopea* 6:813–844.

Cronquist, A. 1997. Angiosperm phylogeny inferred from 18S ribosomal DNA sequences. *Ann. Miss. Bot. Gard.* 84:1–49.

Darwin, C. [1859] 1964. *On the origin of species: A facsimile of the first edition*. Facsimile of 1st ed. ed. Cambridge: Harvard University Press.

Davis, J. I., and K. C. Nixon. 1992. Populations, genetic variation, and the delimitation of phylogenetic species. *Syst. Biol.* 41:421–435.

Dawkins, R. 1976. *The selfish gene*. Oxford: Oxford University Press.

Dawkins, R. 1982. *The extended phenotype*. Oxford: Oxford University Press.

Dawkins, R. 1998. *Unweaving the rainbow*. London: The Penguin Press.

Deacon, T. 1997. *The symbolic species*. New York: Norton.

Dennett, D. C. 1991. Real patterns. *J. Philos.* 88:27–51.

Dennett, D. C. 1993. Back from the drawing board. Pp. 203–235 in *Dennett and his critics*. Edited by B. Dahlbom. Cambridge, MA: Blackwell.

Dennett, D. C. 1995. *Darwin's dangerous idea*. New York: Simon & Schuster.

DePamphilis, C. W., and R. Wyatt. 1989. Hybridization and introgression in buckeyes (Aesculus: Hippocastanaceae): A review of the evidence and a hypothesis to explain long-distance gene. *Syst. Bot.* 14:593–611.

de Queiroz, K. 1996. A phylogenetic approach to biological nomenclature as an alternative to the Linnaean systems in current use. Online: http://www.inform.umd.edu/PBIO/nomcl/dequ.html. *Proc. Mini-Symp. Biol. Nomen. 21st Century*. College Park: University of Maryland. Accessed: March 29, 2001.

de Queiroz, K. 1999. The general lineage concept of species and the defining properties of the species category. Pp. 49–89 in *Species*. Edited by R. A. Wilson. Cambridge: MIT Press.

de Queiroz, K., and M. J. Donoghue. 1988. Phylogenetic systematics and the species problem. *Cladistics* 4:317–338.

de Queiroz, K., and J. Gauthier. 1994. Toward a phylogenetic system of biological nomenclature. *Trans. Ecol. Evol.* 9:27–30.

Diamond, J. M. 1966. Zoological classification system of a primitive people. *Science* 151:1101–1104.

Dobzhansky, T. 1936. Studies of hybrid sterility. II. Localization of sterility factors in *Drosophila pseudoobscura* hybrids. *Genetics* 21:113–135.

Dobzhansky, T. 1937. *Genetics and the origin of species*. 1st ed. New York: Columbia University Press.

Dobzhansky, T. 1940. Speciation as a stage in evolutionary divergence. *Am. Natur.* 74:312–321.

Dobzhansky, T. 1950. Mendelian populations and their evolution. *Am. Natur.* 84:401–418.

Dobzhansky, T. 1951. *Genetics and the origins of species*. 3d ed. New York: Columbia University Press.

Donoghue, M. J. 1985. A critique of the biological species concept and recommendations for a phylogenetic alternative. *Bryologist* 88:172–181.

Dowling, T. E., and B. D. DeMarais. 1993. Evolutionary significance of introgressive hybridization in cyprinid fishes. *Nature* 362:444–446.

Doyle, J. J. 1992. Gene trees and species trees: molecular systematics as one-character taxonomy. *Syst. Bot.* 17:144–163.

Dunn, L. C. 1965. *A short history of genetics*. New York: McGraw-Hill.

Dupré, J. 1999. On the impossibility of a monistic account of species. Pp. 3–22 in *Species*. Edited by R. A. Wilson. Cambridge: MIT Press.

Embley, T. M., and E. Stackebrandt. 1997. Species in practice: Exploring uncultured prokaryote diversity in natural samples. Pp. 61–82 in *Species: the units of biodiversity*. Edited by M. F. Claridge, H. A. Dawah, and M. R. Wilson. London: Chapman & Hall.

Endler, J. A. 1989. Conceptual and other problems in speciation. Pp. 625–648 in *Speciation and its consequences*. Edited by D. Otte and J. A. Endler. Sunderland, MA: Sinauer Associates.

Feldman, M. W., S. P. Otto, and F. B. Christiansen. 1997. Population genetic perspectives on the evolution of recombination. *Annu. Rev. Genet.* 30:261–295.

Felsenstein, J. 1974. The evolutionary advantage of recombination. *Genetics* 78:737–756.

Felsenstein, J. 1975. A pain in the torus: Some difficulties with models of isolation by distance. *Am. Natur.* 109:359–368.

Felsenstein, J. 1981. Skepticism towards Santa Rosalia, or why are there so few kinds of animals. *Evolution* 35:124–138.

Ferris, S. D., R. D. Sage, C. M. Huang, J. T. Nielsen, U. Ritte, and A. C. Wilson.

1983. Flow of mitochondrial DNA across a species boundary. *Proc. Natl. Acad. Sci. USA* 80:2290–2294.

Fields, P. A., and G. N. Somero. 1998. Hot spots in cold adaptation: Localized increases in conformational flexibility in lactate dehydrogenase A4 orthologs of Antarctic notothenioid fishes. *Proc. Natl. Acad. Sci. USA* 95:11476–11481.

Fisher, R. A. 1958. *The genetical theory of natural selection.* 2d ed. New York: Dover.

Fodor, J. and E. Lepore. 1996. The red herring and the pet fish: Why concepts still can't be prototypes. *Cognition* 58:253–270.

Fodor, J. A. 1998. *Concepts: Where cognitive science went wrong.* Oxford: Oxford University Press.

Gallacher, K. G., K. A. Schierenbeck, and C. M. D'Antonio. 1997. Hybridization and introgression in Carpobrotus spp. (Aizoaceae) in California. II. Allozyme evidence. *Am. J. Bot.* 84:905–911.

Garcia, D. K., and S. K. Davis. 1994. Evidence for a mosaic hybrid zone in the grass shrimp *Palaemonetes Kadiakensis* (Palaemonidae) as revealed by multiple genetic markers. *Evolution* 48:376–391.

Garner, K. J., and O. A. Ryder. 1996. Mitochondrial DNA diversity in gorillas. *Mol. Phylogenet. Evol.* 6:39–48.

Gentner, D., and J. Medina. 1998. Similarity and the development of rules. *Cognition* 65:263–297.

Ghiselin, M. T. 1966. On psychologism in the logic of taxonomic controversies. *Syst. Zool.* 15:207–215.

Ghiselin, M. T. 1974. A radical solution to the species problem. *Syst. Zool.* 23:536–544.

Ghiselin, M. T. 1997. *Metaphysics and the origin of species.* Albany: State University of New York Press.

Gibson, R. 1997. Quine on matters ontological. Online: http://www.phil.indiana.edu/ejap/1997.spring/gibson976.html. *Electron. J. Analy. Philos.* 5. Accessed: May 15, 2000.

Gilmour, J. S. L., and J. W. Gregor. 1939. Demes: A suggested terminology. *Nature* 144:333.

Gilmour, J. S. L., and J. Heslop-Harrison. 1954. The deme terminology and the units of micro-evolutionary change. *Genetica* 27:147–161.

Giuffra, E., R. Guyomard, and G. Forneris. 1996. Phylogenetic relationships and introgression patterns between incipient parapatric species of Italian brown trout (*Salmo trutta* L. complex). *Mol. Ecol.* 5:207–220.

Gonzalez, P., and H. A. Lessios. 1999. Evolution of sea urchin retroviral-like (SURL) elements: Evidence from 40 echinoid species. *Mol. Biol. Evol.* 16:938–952.

Goodenough, O. R., and R. Dawkins. 1994. The "St Jude" mind virus. *Nature* 371:23–24.

Goodfellow, M., G. P. Manfio, and J. Chun. 1997. Towards a practical species concept for cultivable bacteria. Pp. 25–59 in *Species: the units of biodiversity.*

Edited by M. F. Claridge, H. A. Dawah, and M. R. Wilson. London: Chapman & Hall.

Goodman, M., W. J. Bailey, K. Hayasaka, M. J. Stanhope, J. Slightom, and J. Czelusniak. 1994. Molecular evidence on primate phylogeny from DNA sequences. *Am. J. Phys. Anthropol.* 94:3–24.

Gorman, B. M. 1983. On the evolution of orbiviruses. *Intervirology* 20:169–80.

Gosselin, L., R. Jobidon, and L. Bernier. 1999. Genetic variability and structure of Canadian populations of *Chondrostereum purpureum*, a potential biophytocide. *Mol. Ecol.* 8:113–122.

Goulson, D., and K. Jerrim. 1997. Maintenance of the species boundary between *Silene dioica* and *S. latifolia* (red and white campion). *Oikos* 79:115–126.

Grant, B. R., and P. R. Grant. 1998. Hybridization and speciation in Darwin's finches. Pp. 404–422 in *Endless forms: species and speciation*. Edited by D. J. Howard and S. H. Berlocher. New York: Oxford University Press.

Grant, P. R., and B. R. Grant. 1997. Genetics and the origin of bird species. *Proc. Natl. Acad. Sci. USA* 94:7768–7775.

Grant, V. 1966. The selective origin of incompatibility barriers in the plant species Gilia. *Am. Natur.* 100:99–118.

Grant, V. 1981. *Plant speciation*. 2d ed. New York: Columbia University Press.

Green, C. D., and J. Vervaeke. 1997. The experience of objects and the objects of experience. *Metaphor Symbol* 12:3–17.

Gregg, J. R. 1950. Taxonomy, language and reality. *Am. Natur.* 84:419–435.

Gueiros-filho, F. J., and S. M. Beverley. 1997. Trans-kingdom transposition of the Drosophila element mariner within the protozoan Leishmania. *Science* 276:1716–1719.

Gupta, R. C., L. R. Bazemore, E. I. Golub, and C. M. Radding. 1997. Activities of human recombination protein Rad51. *Proc. Natl. Acad. Sci. USA* 94:463–468.

Gupta, R. S., and G. B. Golding. 1996. The origin of the eukaryotic cell. *Trends. Bio. Sci.* 21:166–171.

Gupta, R. S., and B. Singh. 1994. Phylogenetic analysis of 70 kD heat shock protein sequences suggests a chimeric origin for the eukaryotic cell nucleus. *Curr. Biol.* 4:1104–1114.

Gyllensten, U. B., D. Lashkari, and H. A. Erlich. 1990. Allelic diversification at the class II DQB locus of the mammalian major histocompatibility complex. *Proc. Natl. Acad. Sci. USA* 87:1835–1839.

Hacking, I. 1983. *Representing and intervening: Introductory topics in the philosophy of natural science*. New York: Cambridge University Press.

Hagen, R. H., and J. M. Scriber. 1989. Sex-linked diapause, color and allozyme loci in *Papilio glaucus*.: Linkage analysis and significance in hybrid zone. *J. Heredity* 80:179–185.

Hagen, R. H., and J. M. Scriber. 1995. Sex chromosomes and speciation in Swallowtail butterflies. Pp. 211–227 in *Swallowtail butterfiles: Their ecology &*

evolutionary biology. Edited by J. M. Scriber, Y. Tsubaki, and R. C. Lederhouse. Gainesville, FL: Scientific Publishers.

Hahn, U., and N. Chater. 1998. Similarity and rules: Distinct? exhaustive? empirically distinguishable? *Cognition* 65:197–230.

Hampton, J. A. 1995. Testing the prototype theory of concepts. *J. Memory Language* 34:686–708.

Harrison, R. C. 1990. Hybrid zones: windows on an evolutionary process. Pp. 69–128 in *Oxford Surveys in Evolutionary Biology*, vol. 7. Edited by D. Futuyma and J. Antonovics. New York: Oxford University Press.

Harrison, R. G., and S. M. Bogdanowicz. 1997. Patterns of variation and linkage disequilibrium in a field cricket hybrid zone. *Evolution* 51:493–505.

Harrison, R. G., and D. M. Rand. 1989. Mosaic hybrid zones and the nature of species boundaries. Pp. 111–133 in *Speciation and its consequences*. Edited by D. Otte and J. A. Endler. Sunderland, MA: Sinauer Associates.

Harvey, P. H., and M. D. Pagel. 1991. *The comparative method in evolutionary biology*. Oxford: Oxford University Press.

Haugeland, J. 1993. Pattern and being. Pp. 53–69 in *Dennett and his critics*. Edited by B. Dahlbom. Cambridge, MA: Blackwell.

Hawksworth, D. L., R. E. Ricklefs, R. M. Cowling, and M. J. Samways. 1995. Magnitude and distribution of biodiversity. Pp. 107–191 in *Global biodiversity assessment*. Edited by V. H. Heywood. Cambridge: Cambridge University Press.

Heath, D. D., P. D. Rawson, and T. J. Hilbish. 1995. PCR-based nuclear markers identify alien blue mussel (Mytilus spp.) genotypes on the west coast of Canada. *Canadian J. Fisheries Aquatic Sci.* 52:2621–2627.

Hendrix, R. W., M. C. Smith, R. N. Burns, M. E. Ford, and G. F. Hatfull. 1999. Evolutionary relationships among diverse bacteriophages and prophages: All the world's a phage. *Proc. Natl. Acad. Sci. USA* 96:2192–2197.

Hennig, W. 1966. *Phylogenetic systematics*. Urbana: University of Illinois Press.

Hey, J. 1994. Bridging phylogenetics and population genetics with gene tree models. Pp. 435–449 in *Molecular approaches to ecology and evolution*. Edited by B. Schierwater, B. Streit, G. Wagner, and R. DeSalle. Basel, Germany: Birkhäuser-Verlag.

Hey, J. 1998. Selfish genes, pleiotropy and the origin of recombination. *Genetics* 149:2089–2097.

Hey, J., and R. M. Kliman. 1993. Population genetics and phylogenetics of DNA sequence variation at multiple loci within the *Drosophila melanogaster* species complex. *Mol. Biol. Evol.* 10:804–822.

Heywood, V. H. 1998. The species concept as a socio-cultural phenomenon—a source of the scientific dilemma. *Theory Biosci.* 117:203–212.

Heywood, V. H., and R. T. Watson, eds. 1995. *Global biodiversity assessment*. Cambridge: Cambridge University Press.

Hibbett, D. S., and M. J. Donoghue. 1998. Integrating phylogenetic analysis and classification in fungi. *Mycologia* 90:347–356.

Hill, W. G., and A. Robertson. 1966. The effect of linkage on limits to artificial selection. *Genet. Res. Camb.* 8:269–294.

Hilton, H., and J. Hey. 1997. A multilocus view of speciation in the *Drosophila virilis* group reveals complex histories and taxonomic conflicts. *Genet. Res. Camb.* 70:185–194.

Hilton, H., R. M. Kliman, and J. Hey. 1994. Using hitchhiking genes to study adaptation and divergence during speciation within the *Drosophila melanogaster* complex. *Evolution* 48:1900–1913.

Hollocher, H., C. T. Ting, F. Pollack, and C. I. Wu. 1997a. Incipient speciation by sexual isolation in *Drosophila melanogaster*: Variation in mating preference and correlation between sexes. *Evolution* 51:1175–1181.

Hollocher, H., C. T. Ting, M. L. Wu, and C. I. Wu. 1997b. Incipient speciation by sexual isolation in *Drosophila melanogaster*: Extensive genetic divergence without reinforcement. *Genetics* 147:1191–1201.

Howard, D. J., and S. H. Berlocher, eds. 1998. *Endless forms: Species and speciation.* New York: Oxford University Press.

Hudson, R. R., and N. L. Kaplan. 1985. Statistical properties of the number of recombination events in the history of a sample of DNA sequences. *Genetics* 111:147–164.

Hull, D. L. 1965a. The effect of essentialism on taxonomy—two thousand years of stasis (I). *Brit. J. Phil. Sci.* 15:314–326.

Hull, D. L. 1965b. The effect of essentialism on taxonomy—two thousand years of stasis (II). *Brit. J. Phil. Sci.* 16:1–18.

Hull, D. L. 1976. Are species really individuals. *Syst. Zool.* 15:174–191.

Hull, D. L. 1978. A matter of individuality. *Phil. Sci.* 45:335–360.

Hull, D. L. 1988. *Science as a process: An evolutionary account of the social and conceptual development of science.* Chicago, IL: University of Chicago Press.

Hull, D. L. 1997. The ideal species concept—and why we cannot get it. Pp. 357–380 in *Species: The units of biodiversity.* Edited by M. F. Claridge, H. A. Dawah, and M. R. Wilson. London: Chapman & Hall.

Hull, D. L. 1999. On the plurality of species: Questioning the party line. Pp. 23–48 in *Species.* Edited by R. A. Wilson. Cambridge: MIT Press.

Humphries, C. J., P. H. Williams, and R. I. Vane-Wright. 1995. Measuring biodiversity value for conservation. *Ann. Rev. Ecol. Syst.* 26:93–111.

Hunt, D. J. 1997. Nematode species: Concepts and identification strategies exemplified by the Longidoridae, Steinernematidae and Heterorhabditidae. Pp. 221–246 in *Species: The units of biodiversity.* Edited by M. F. Claridge, H. A. Dawah, and M. R. Wilson. London: Chapman & Hall.

Hurford, J. R., M. Studdert-Kennedy, and C. Knight, eds. 1998. *Approaches to the evolution of language: Social and cognitive bases.* Cambridge: Cambridge University Press.

Hutchinson, G. E. 1958. Concluding remarks. *Cold Spring Harbor Symp. Quant. Bio.* 22:415–427.

Huxley, J. S. 1940. Towards the new systematics. Pp. 1–46 in *The new systematics*. Edited by J. S. Huxley. Oxford: Oxford University Press.

Jain, R., M. C. Rivera, and J. A. Lake. 1999. Horizontal gene transfer among genomes: The complexity hypothesis. *Proc. Natl. Acad. Sci. USA* 96:3801–3806.

John, D. M., and C. A. Maggs. 1997. Species problems in eukaryotic algae: A modern perspective. Pp. 83–108 in *Species: The units of biodiversity*. Edited by M. F. Claridge, H. A. Dawah, and M. R. Wilson. London: Chapman & Hall.

Kaessmann, H., V. Wiebe, and S. Pääbo. 1999. Extensive nuclear DNA sequence diversity among chimpanzees. *Science* 286:1159–1162.

Kamau, L., W. A. Hawley, T. Lehmann, A. S. Orago, A. Cornel, Z. Ke, and F. H. Collins. 1998. Use of short tandem repeats for the analysis of genetic variability in sympatric populations of *Anopheles gambiae* and *Anopheles arabiensis*. *Heredity* 80:675–82.

Kaplan, N., R. R. Hudson, and C. H. Langley. 1989. The "hitchhiking effect" revisited. *Genetics* 123:887–899.

Keller, E. F., and E. A. Lloyd. 1992. Introduction. Pp. 1–6 in *Keywords in evolutionary biology*. Edited by E. F. Keller and E. A. Lloyd. Cambridge: Harvard University Press.

Kimura, M., and T. Maruyama. 1966. The mutational load with epistatic gene interactions in fitness. *Genetics* 54:1337–1351.

Klein, J., N. Takahata, and F. J. Ayala. 1993. MHC polymorphism and human origins. *Sci. Am.* 269:78–83.

Kliman, R. M., and J. Hey. 1993. DNA sequence variation at the period locus within and among species of the *Drosophila melanogaster* complex. *Genetics* 133:375–387.

Kondrashov, A. S. 1984. Deleterious mutations as an evolutionary factor. 1. The advantage of recombination. *Genet. Res. Camb.* 44:199–217.

Kondrashov, A. S. 1988. Deleterious mutations and the evolution of sexual reproduction. *Nature* 336:435–440.

Kondrashov, A. S. 1993. Classification of hypotheses on the advantage of amphimixis. *J. Hered.* 84:372–387.

Konings, A., and M. Geerts. 1999. Comments on the revision of *Pseudotropheus zebra*. *Cichlid News* 8:18–26.

Lakoff, G. 1987. *Women, fire, and dangerous things: What categories reveal about the mind*. Chicago, IL: University of Chicago Press.

Landesman, C. 1971. Introduction. Pp. 3–17 in *The problem of universals*. Edited by C. Landesman. New York: Basic Books.

Lane, R. 1997. The species concept in blood-sucking vectors of human diseases. Pp. 273–290 in *Species: The units of biodiversity*. Edited by M. F. Claridge, H. A. Dawah, and M. R. Wilson. London: Chapman & Hall.

Lanzaro, G. C., Y. T. Toure, J. Carnahan, L. Zheng, G. Dolo, S. Traore, V. Petrarca,

K. D. Vernick, and C. E. Taylor. 1998. Complexities in the genetic structure of *Anopheles gambiae* populations in west Africa as revealed by microsatellite DNA analysis. *Proc. Natl. Acad. Sci. USA* 95:14260–14265.

Lawrence, J. G., and H. Ochman. 1998. Molecular archaeology of the *Escherichia coli* genome. *Proc. Natl. Acad. Sci. USA* 95:9413–9417.

Lenski, R. E. 1993. Assessing the genetic structure of microbial populations. *Proc. Natl. Acad. Sci. USA* 90:4334–4336.

Levin, D. A. 1979. The nature of plant species. *Science* 204:381–384.

L'Haridon, S., A. L. Reysenbach, P. Glenat, D. Prieur, and C. Jeanthon. 1995. Hot subterranean biosphere in a continental oil reservoir. *Nature* 377:223–224.

Lopez-Garc, P., and D. Moreira. 1999. Metabolic symbiosis at the origin of eukaryotes. *Trends Biochem. Sci.* 24:88–93.

Majewski, J., and F. M. Cohan. 1998. The effect of mismatch repair and heteroduplex formation on sexual isolation in Bacillus. *Genetics* 148:13–18.

Mallet, J. 1995. A species definition for the modern synthesis. *Trans. Ecol. Evol.* 10:294–299.

Mandelbrot, B. B. 1977. *Fractals: Form, chance & dimension.* San Francisco, CA: Freeman.

Margulis, L., and D. Sagan. 1986. *Origins of sex: Three billion years of genetic recombination.* New Haven, CT: Yale University Press.

Martin, G. 1996. Birds in double trouble. *Nature* 380:666–667.

Martin, J., E. Herniou, J. Cook, R. W. O'Neill, and M. Tristem. 1999. Interclass transmission and phyletic host tracking in murine leukemia virus-related retroviruses. *J. Virol.* 73:2442–2449.

Martinich, A. P. 1996. *The philosophy of language.* 3d ed. New York: Oxford University Press.

Masterson, J. 1994. Stomatal size in fossil plants: Evidence for polyploidy in majority of angiosperms. *Science* 264:421–424.

Matic, I., C. Rayssiguier, and M. Radman. 1995. Interspecies gene exchange in bacteria: The role of SOS and mismatch repair systems in evolution of species. *Cell* 80:507–515.

Mayden, R. L. 1997. A hierarchy of species concepts: The denouement in the saga of the species problem. Pp. 381–424 in *Species: The units of biodiversity.* Edited by M. F. Claridge, H. A. Dawah, and M. R. Wilson. London: Chapman & Hall.

Maynard Smith, J. 1986. *The problems of biology.* Oxford: Oxford University Press.

Maynard Smith, J., and J. Haigh. 1974. The hitch-hiking effect of a favourable gene. *Genet. Res. Camb.* 23:23–35.

Maynard Smith, J., and E. Szathmáry. 1995. *The major transitions of evolution.* Oxford: W. H. Freeman/Spectrum.

Maynard Smith, J., and E. Szathmáry. 1999. *The origins of life: From the birth of life to the origin of language.* Oxford: Oxford University Press.

Mayr, E. 1942. *Systematics and the origin of species.* New York: Columbia University Press.

Mayr, E. 1963. *Animal species and evolution.* Cambridge: Belknap Press of Harvard University Press.

Mayr, E. 1982. *The growth of biological thought.* Cambridge: Harvard University Press.

Mayr, E. 1987. The ontological status of species: Scientific progress and philosophical terminology. *Biol. Phil.* 2:145–166.

Mayr, E. 1992. A local flora and the biological species concept. *Am. J. Bot.* 79:222–238.

Mayr, E. 1996. What is a species and what is not? *Phil. Sci.* 63:262–277.

Mayr, E., E. G. Linsley, and R. L. Usinger. 1953. *Methods and principles of systematic zoology.* New York: McGraw-Hill.

Medin, D., and S. Atran. 1999. *Folkbiology.* Cambridge: MIT Press.

Medin, D. L., and M. M. Schaeffer. 1978. Context theory of classification learning. *Psych. Rev.* 85:207–238.

Mishler, B. D., and M. J. Donoghue. 1982. Species concepts: A case for pluralism. *Syst. Zool.* 31:491–503.

Mishler, B. D., and E. C. Theriot. 2000. The phylogenetic species concept (*sensu* Mishler and Theriot): Monophyly, apomorphy and phylogenetic species concepts. Pp. 44–54 in *Species concepts and phylogenetic theory: A debate.* Edited by Q. D. Wheeler and R. Meier. New York: Columbia University Press.

Morin, P. A., J. J. Moore, and D. S. Woodruff. 1992. Identification of chimpanzee subspecies with DNA from hair and allele-specific probes. *Proc. R. Soc. Lond. B Biol. Sci.* 249:293–297.

Moritz, R. F. A., and E. E. Southwick. 1992. *Bees as superorganisms: An evolutionary reality.* New York: Springer-Verlag.

Muller, H. J. 1929. The gene as the basis of life. *Proc. Intern. Cong. Plant Sci.* 1:897–921.

Munson, M. A., D. B. Nedwell, and T. M. Embley. 1997. Phylogenetic diversity of Archaea in sediment samples from a coastal salt marsh. *Appl. Environ. Microbiol.* 63:4729–4733.

Nelson, G. 1989. Species and taxa: Systematics and evolution. Pp. 60–81 in *Speciation and its consequences.* Edited by D. Otte and J. A. Endler. Sunderland, MA: Sinauer Associates.

Newmeyer, F. J. 1998. On the supposed "counterfunctionality" of universal grammar: Some evolutionary implications. Pp. 305–319 in *Approaches to the evolution of language.* Edited by J. R. Hurford, M. Studdert-Kennedy, and C. Knight. Cambridge: Cambridge University Press.

Nixon, K. C., and Q. D. Wheeler. 1990. An amplification of the phylogenetic species concept. *Cladistics* 6:211–223.

Noor, M. A. 1995. Speciation driven by natural selection in Drosophila. *Nature* 375:674–675.

Nosofsky, R. M. 1984. Choice, similarity, and the context theory of classification. *J. Exper. Psych.: Learning Mem. Cognit.* 10:104–114.

Ochman, H., and U. Bergthorsson. 1998. Rates and patterns of chromosome evolution in enteric bacteria. *Curr. Opin. Microb.* 1:580–583.

O'Connor, J. J., and E. F. Robertson. 2000. Niels Fabian Helge von Koch. Online: http://www-history.mcs.st-andrews.ac.uk/history/Mathematicians/Koch.html. School of Mathematics and Statistics, University of St. Andrews, Scotland. Accessed: March 29, 2001.

Orgel, L. E. 1992. Molecular replication. *Nature* 358:203–209.

Osherson, D. N., and E. E. Smith. 1981. On the adequacy of prototype theory as a theory of concepts. *Cognition* 9:35–58.

Otto, S. P., and N. H. Barton. 1997. The evolution of recombination: Removing the limits to natural selection. *Genetics* 147:879–906.

Packer, L., and J. S. Taylor. 1997. How many hidden species are there? An application of the phylogenetic species concept to genetic data for some comparatively well known bee "species." *Canad. Entomol.* 129:587–594.

Pamilo, P., and M. Nei. 1988. Relationships between gene trees and species trees. *Mol. Biol. Evol.* 5:568–583.

Peterson, A. T., and A. G. Navarro-Siguenza. 1999. Alternate species concepts as bases for determining priority conservation areas. *Conserv. Biol.* 13:427–431.

Piel, W. H., and K. J. Nutt. 2000. One species or several? Discordant patterns of geographic variation between allozymes and mtDNA sequences among spiders in the Genus Metepeira (Araneae: Araneidae). *Mol. Phylogenet. Evol.* 15:414–418.

Pinker, S. 1994. *The lanugage instinct.* New York: W. Morrow.

Pinker, S., and P. Bloom. 1990. Natural language and natural selection. *Behav. Brain Sci.* 13:707–784.

Platnick, N. I. 1979. Philosophy and the transformation of cladistics. *Syst. Zool.* 28:537–546.

Popper, K. R. 1959. *The logic of scientific discovery.* New York: Harper & Row.

Porter, A., and G. Ribi. 1994. Population genetics of Viviparus (Mollusca: Prosobranchia): Homogeneity of *V. ater* and apparent introgression into *V. contectus. Heredity* 73:170–176.

Powell, J. R. 1983. Interspecific cytoplasmic gene flow in the absence of nuclear gene flow: Evidence from Drosophila. *Proc. Natl. Acad. Sci. USA* 80:492–495.

Provine, W. B. 1971. *The origins of theoretical population genetics.* Chicago, IL: University of Chicago Press.

Purvis, O. W, 1997. The species concept in lichens. Pp. 109–134 in *Species: The units of biodiversity.* Edited by M. F. Claridge, H. A. Dawah, and M. R. Wilson. London: Chapman & Hall.

Putnam, H. 1981. *Reason, truth and history.* Cambridge: Cambridge University Press.

Quine, W. V. O. 1960. *Word & Object*. New York: Wiley.

Quine, W. V. O. 1961. On what there is. Pp. 1–19 in *From a logical point of view*. 2d ed. Edited by W. V. O. Quine. Cambridge: Harvard University Press.

Ravin, A. W. 1963. Experimental approaches to the study of bacterial phylogeny. *Am. Natur.* 97:307–318.

Rieseberg, L. H., B. Sinervo, C. R. Linder, M. C. Ungerer, and D. M. Arias. 1996. Role of gene interactions in hybrid speciation: Evidence from ancient and experimental hybrids. *Science* 272:741–745.

Rieseberg, L. H., and J. F. Wendel. 1993. Introgression and its consequences in plants. Pp. 70–109 in *Hybrid zones and the evolutionary process*. Edited by R. G. Harrison. New York: Oxford University Press.

Rieseberg, L. H., J. Whitton, and C. R. Linder. 1996. Molecular marker incongruence in plant hybrid zones and phylogenetic trees. *Acta Botanica Neerlandica* 45:243–262.

Robertson, H. M., and D. J. Lampe. 1995. Distribution of transposable elements in arthropods. *Annu. Rev. Entomol.* 40:333–57.

Rojas, M. 1992. The species problem and conservation: What are we protecting. *Conserv. Biol.* 6:170–178.

Rosch, E. 1978. Principles of categorization. Pp. 28–48 in *Cognition and categorization*. Edited by E. Rosch and B. B. Lloyd. Hillsdale, NJ: Erlbaum.

Rosch, E., C. Simpson, and R. S. Miller. 1976. Structural basis of typicality effects. *J. Exper. Psych.: Human Percep. Perform.* 2:491–502.

Rosen, D. E. 1979. Fishes from the uplands and intermontane basins of Guatemala: Revisionary studies and comparative biogeography. *Bull. Am. Mus. Nat. Hist.* 162:267–376.

Ross, K. G., D. D. Shoemaker, M. J. Krieger, C. J. DeHeer, and L. Keller. 1999. Assessing genetic structure with multiple classes of molecular markers: A case study involving the introduced fire ant *Solenopsis invicta. Mol. Biol. Evol.* 16:525–543.

Ross, W. D. 1964. *Aristotle*. 5th ed. London: Methuen.

Rundle, H. D., and D. Schluter. 1998. Reinforcement of stickleback mate preferences: Sympatry breeds contempt. *Evolution* 52:200–208.

Russo, C. A., N. Takezaki, and M. Nei. 1995. Molecular phylogeny and divergence times of Drosophilid species. *Mol. Biol. Evol.* 12:391–404.

Ruvolo, M. 1997. Molecular phylogeny of the Hominoids: Inferences from multiple independent DNA sequence data sets. *Mol. Biol. Evol.* 14:248–265.

Samuel, R., W. Pinsker, and F. Ehrendorfer. 1995. Electrophoretic analysis of genetic variation within and between populations of *Quercus cerris, Q. pubescens, Q. petraea* and *Q. robur* (Fagaceae) from eastern Austria. *Botanica Acta.* 108:290–299.

Sanderson, N. 1989. Can gene flow prevent reinforcement? *Evolution* 43:1223–1235.

Savage-Rumbaugh, S., S. G. Shanker, and T. J. Taylor. 1998. *Apes, language and the human mind*. New York: Oxford University Press.

208   REFERENCES

Shaw, D. D., P. Wilkinson, and C. Moran. 1979. A comparison of chromosomal and allozymal variation across a narrow hybrid zone in the grasshopper *Caledia captiva. Chromosoma* 75:333–351.

Shinohara, A., H. Ogawa, Y. Matsuda, N. Ushio, K. Ikeo, and T. Ogawa. 1993. Cloning of human, mouse and fission yeast recombination genes homologous to RAD51 and recA. *Nat. Genet.* 4:239–243.

Simpson, G. G. 1961. *Principles of animal taxonomy.* New York: Columbia University Press.

Singer, C. 1950. *A short history of biology,* 2d ed. New York: Henry Schuman.

Sites, J. W., and K. A. Crandall. 1997. Testing species boundaries in biodiversity studies. *Conserv. Biol.* 11:1289–1297.

Smith, E. E., and D. L. Medin. 1981. *Categories and concepts.* Cambridge: Harvard University Press.

Smith, E. E., A. L. Patalano, and J. Jonides. 1998. Alternative strategies of categorization. *Cognition* 65:167–196.

Sokal, R. R., and T. J. Crovello. 1970. The biological species concept: A critical evaluation. *Am. Natur.* 104:127–153.

Sokal, R. R., and P. H. A. Sneath. 1963. *Principles of numerical taxonomy.* San Francisco, CA: Freeman.

Somero, G. N., and P. S. Low. 1976. Temperature: A "shaping force" in protein evolution. *Biochem. Soc. Symp.* 41:33–42.

Stamos, D. N. 1996. Was Darwin really a species nominalist? *J. Hist. Biol.* 29:127–144.

Stauffer, J. R. J., N. J. Bowers, K. A. Kellogg, and K. R. McKaye. 1997. A revision of the blue-black *Pseudotropheus zebra* complex from Lake Malawi, Africa, with a description of a new genus and ten new species. *Proc. Acad. Sci. Philadelphia* 148:189–230.

Stebbins, G. L. 1950. *Variation and evolution in plants.* New York: Columbia University Press.

Stebbins, G. L. 1969. Comments on the search for a "perfect system." *Taxon* 18:357–359.

Stevens, P. F. 1990. Species: Historical perspectives. Pp. 302–311 in *Keywords in evolutionary biology.* Edited by E. F. Keller and E. Lloyd. Cambridge: Harvard University Press.

Sung, P. 1994. Catalysis of ATP-dependent homologous DNA pairing and strand exchange by yeast RAD51 protein. *Science* 265:1241–1243.

Tajima, F. 1983. Evolutionary relationships of DNA sequences in finite populations. *Genetics* 105:437–460.

Takahata, N., and Y. Satta. 1997. Evolution of the primate lineage leading to modern humans: Phylogenetic and demographic inferences from DNA sequences. *Proc. Natl. Acad. Sci. USA* 94:4811–4815.

Taylor, D. J., and P. D. Hebert. 1993. Habitat-dependent hybrid parentage and differential introgression between neighboring sympatric Daphnia species. *Proc. Natl. Acad. Sci. USA* 90:7079–7083.

Templeton, A. 1998. Species and speciation: Geography, population structure, ecology and gene trees. Pp. 32–43 in *Endless forms: Species and speciation*. Edited by D. J. Howard and S. H. Berlocher. New York: Oxford University Press.

Templeton, A. R. 1989. The meaning of species and speciation: A genetic perspective. Pp. 3–27 in *Speciation and its consequences*. Edited by D. Otte and J. A. Endler. Sunderland, MA: Sinauer Associates.

Templeton, A. R. 1994. The role of molecular genetics in speciation studies in *Molecular approaches to ecology and evolution*. Edited by B. Schierwater, B. Streit, G. Wagner, and R. DeSalle. Basel, Germany: Birkhäuser-Verlag.

Thagard, P. 1996. The concept of disease: structure and change. *Communication and Cognition* 29:445–478.

Ungerer, F., and H.-J. Schmid. 1996. *An introduction to cognitive linguistics*. London: Longman.

Van Regenmortel, M. H. V. 1997. Viral species. Pp. 17–24 in *Species: The units of Biodiversity*. Edited by M. F. Claridge, H. A. Dawah, and M. R. Wilson. London: Chapman & Hall.

Van Valen, L. 1976. Ecological species, multispecies, and oaks. *Taxon* 25:233–239.

Vane-Wright, R. I. 1996. Identifying priorities for the conservation of biodiversity: Systematic biological criteria within a socio-political framework. Pp. 309–344 in *Biodiversity: A biology of numbers and difference*. Edited by K. J. Gaston. Oxford: Blackwell.

Vane-Wright, R. I., C. J. Humphries, and P. H. Williams. 1991. What to protect?—systematics and the agony of choice. *Biol. Conserv.* 55:235–254.

Vanlerberghe, F., P. Boursot, J. Catalan, S. Gerasimov, F. Bonhomme, B. A. Botev, and L. Thaler. 1988. Genetic analysis of the hybridization zone between two subspecies *Mus musculus domesticus* and *Mus musculus musculus* in Bulgaria. *Genome* 30:427–437.

Vervaeke, J., and C. D. Green. 1997. Women, fire, and dangerous theories: A critique of Lakoff's theory of categorization. *Metaphor Symbol.* 12:59–80.

Vetriani, C., A. L. Reysenbach, and J. Dore. 1998. Recovery and phylogenetic analysis of archaeal rRNA sequences from continental shelf sediments. *FEMS Microbiol. Lets.* 161:83–88.

Vila, C., J. E. Maldonado, and R. K. Wayne. 1999. Phylogenetic relationships, evolution, and genetic diversity of the domestic dog. *J. Hered.* 90:71–77.

Vila, C., P. Savolainen, J. E. Maldonado, I. R. Amorim, J. E. Rice, R. L. Honeycutt, K. A. Crandall, J. Lundeberg, and R. K. Wayne. 1997. Multiple and ancient origins of the domestic dog. *Science* 276:1687–1689.

Vrana, P., and W. Wheeler. 1992. Individual organisms as terminal entities: Laying the species problem to rest. *Cladistics* 8:67–72.

Wagner, W. H. J. 1970. Biosystematics and evolutionary noise. *Taxon* 19:146–151.

Wakeley, J., and J. Hey. 1997. Estimating ancestral population parameters. *Genetics* 145:847–855.

Wakeley, J., and J. Hey. 1998. Testing speciation models with DNA sequence data. Pp. 157–175 in *Molecular approaches to ecology and evolution*. Edited by R. DeSalle and B. Schierwater. Basel, Germany: Birkhäuser-Verlag.

Wallace, A. R. 1901. *Darwinism*, 3d ed. London: Macmillan.

Wang, R. L., J. Wakeley, and J. Hey. 1997. Gene flow and natural selection in the origin of *Drosophila pseudoobscura* and close relatives. *Genetics* 147:1091–1106.

Warwick, R. M., and K. R. Clarke. 1998. Taxonomic distinctness and environmental assessment. *J. Appl. Ecol.* 35:532–543.

Watterson, G. A. 1982. Substitution times for mutant nucleotides. *J. Appl. Prob.* 19A:59–70.

Watzlawick, P., J. B. Bavelas, and D. D. Jackson. 1967. *Pragmatics of human communication: A study of interactional patterns, pathologies and paradoxes.* New York: Norton.

Wendel, J. F., J. M. Stewart, and J. H. Rettig. 1991. Molecular evidence for homoploid reticulate evolution among Australian species of Gossypium. *Evolution* 45:694–711.

Wheeler, Q. D., and R. Meier, eds. 2000. *Species concepts and phylogenetic theory: A debate.* New York: Columbia University Press.

Wheeler, Q. D., and N. I. Platnick. 2000. The phylogenetic species concept (*sensu* Wheeler and Platnick). Pp. 55–69 in *Species concepts and phylogenetic theory: A debate.* Edited by Q. D. Wheeler and R. Meier. New York: Columbia University Press.

Whittemore, A. T. 1993. Species concepts: A reply to Ernst Mayr. *Taxon* 42:573–583.

Wiley, E. O. 1981. *Phylogenetics: The theory and practice of phylogenetic systematics.* New York: Wiley.

Williams, P. H., and K. J. Gaston. 1994. Measuring more of biodiversity: Can higher-taxon richness predict wholesale species richness? *Biol. Conserv.* 67:211–217.

Williams, P. H., K. J. Gaston, and C. J. Humphries. 1997. Mapping biodiversity value worldwide: Combining higher-taxon richness from different groups. *Proc. R. Soc. Lond. B Biol. Sci.* 264:141–148.

Wilson, J. 1999. *Biological individuality: The identity and persistence of living entities.* Cambridge: Cambridge University Press.

Winston, J. 1999. *Describing species.* New York: Columbia University Press.

Wittgenstein, L. 1968. *Philosophical investigations.* New York: Macmillan.

Wright, S. 1931. Evolution in Mendelian populations. *Genetics* 16:97–159.

Wright, S. 1943. Isolation by distance. *Genetics* 28:114–138.

Wu, C. I., H. Hollocher, D. J. Begun, C. F. Aquadro, Y. Xu, and M. L. Wu. 1995. Sexual isolation in *Drosophila melanogaster*: A possible case of incipient speciation. *Proc. Natl. Acad. Sci. USA* 92:2519–2523.

Závodszky, P., J. Kardos Á. Svingor, and G. A. Petsko. 1998. Adjustment of conformational flexibility is a key event in the thermal adaptation of proteins. *Proc. Natl. Acad. Sci. USA* 95:7406–7411.

Zawadzki, P., M. S. Roberts, and F. M. Cohan. 1995. The log-linear relationship between sexual isolation and sequence divergence in Bacillus transformation is robust. *Genetics* 140:917–932.

Zink, R. M. 1996. Bird species diversity. *Nature* 381:566.

Zirkle, C. 1959. Species before Darwin. *Proc. Am. Philos. Soc.* 103:636–644.

INDEX

acellular slime molds, 33, 36
adaptation, 74–76, 80–81, 89,
   103–105, 111, 118–128, 175
adjectives, 48, 58, 170
agamospecies, 11
anthropology. *See* taxa
Aristotle, 6, 49, 61, 64, 65,106

bacteria, 30, 31, 93, 100, 141, 142, 154
Barton, N., 90, 91
Berlin, B., 63, 64, 65
biological diversity
   basic causes of, 13
   fact of, 69, 133
   preservation of, 189–191
biological species concept. *See*
   species concepts
birds, 8, 20, 54, 58, 94, 99,100, 176, 177
boundaries
   of categories. *See* categories
   of entities in nature, 22, 37, 38, 47–
      48, 79–80, 116, 157, 160, 181
   of evolutionary groups. *See* evolu-
      tionary groups
   perception of, 112–113
   of prototypes. *See* prototype ef-
      fects
   of species. *See* species
Briggs, D., 11, 12

categories
   Aristotelian, 49, 54, 64
   boundaries of, 55, 59, 115–116,
      124
   classical view of, 49, 54, 64
   combined to make concepts, 57
   in the mind, 49, 51–53, 111,
      114–116, 128
   natural kinds, 50, 51–53, 55, 59,
      62, 68, 98, 105–106, 111–117,
      161–162, 173–176
   as predicates, 46, 59
   purpose of, 48–49
   as rules, 54–59, 115–116, 121,
      124, 153
   See also prototype effects
cell, definition of term, 36
Chomsky, N., 122
cladistics, 147–149, 155
clouds, 22–23, 47, 48, 101, 156, 186
cohesion species concept. *See* species
   concepts
common structure model, 46, 52
competition, 70, 74, 75, 76, 80–81,
   85, 86, 87, 97–99, 103, 105
concept conflict, 18–20, 23, 24
count creep, 18, 20, 24, 161
Crick, F., 25, 119
cytospecies, 11

213